# Sociological Justice

# Sociological Justice

## DONALD BLACK

New York   Oxford
OXFORD UNIVERSITY PRESS
1989

Oxford University Press

Oxford   New York   Toronto
Delhi   Bombay   Calcutta   Madras   Karachi
Petaling Jaya   Singapore   Hong Kong   Tokyo
Nairobi   Dar es Salaam   Cape Town
Melbourne   Auckland

and associated companies in
Berlin   Ibadan

Published by Oxford University Press, Inc.,
200 Madison Avenue, New York, New York 10016

Library of Congress Cataloging-in-Publication Data
Black, Donald J.
Sociological justice / Donald Black.
p.   cm.
Bibliography: p.
Includes index
ISBN 0-19-505447-4
1. Sociological jurisprudence.   2. Justice, Administration of.
I. Title.
K370.B55   1989
340'.115—dc 19   88-17682   CIP

2 4 6 8 9 7 5 3 1
Printed in the United States of America
on acid-free paper

To
M. P.

# Preface

A new branch of legal sociology addresses how the social structure of cases predicts the way they are handled. Who has a complaint against whom? Who supports each side? Who decides the result? We can specify the social elevation of each party, the social distance between them, whether they are individual or corporate beings, and so on. These considerations are fateful. They tell us, for example, who is likely to win.

The following pages explore the significance of this field—the sociology of the case—for law itself. Sociological knowledge has applications in the practice of law, in legal reform, and in jurisprudence and social policy. Lawyers can use it to advance their interests (such as their chances of winning cases), reformers to reduce discrimination (such as by socially homogenizing the cases), and legal scholars and policymakers to assess the reality and morality of law (such as the ideal of universalistic treatment).

Law is more than rules and logic. It is also people. It cannot be understood in a social vacuum. Even so, recent developments in legal sociology are largely unknown and unexploited.

The title of this book, *Sociological Justice*, refers to the self-conscious application of sociology to legal action. This is now beginning, and it may change law forever.

The ideas here were largely conceived and developed at Harvard Law School during the years 1979–1985, and delivered in lectures for my course called "Sociology of Law." A number of students offered helpful suggestions and elaborations. Their response encouraged me to seek a wider audience.

Administrative support was provided by Harvard Law School's Center for Criminal Justice, directed over the years by James Vorenberg (now Dean), Lloyd E. Ohlin, Philip Heymann, the late Alden D. Miller (Associate Director), and Daniel McGillis (Deputy Director). I completed the book at the University of Virginia's Department of Sociology. Financial support was partially provided by the National Science Foundation Program in Law and Social Science.

M. P. Baumgartner contributed to this book in many ways during years of discussion and commented extensively on an earlier draft. Others also improved the manuscript: Valerie Aubry, Piers Beirne, Theodore Caplow, Mark Cooney, John Herrmann (who prepared the index as well), Allan Horwitz, Calvin Morrill, Marion Osmun, Roberta Senechal, James Tucker, Rosemary Wellner, and Eliot Werner. Joan Snapp expertly typed every draft.

I am grateful to all these individuals and organizations.

*Free Union, Virginia*                                          D. B.
*May 1988*

# Contents

# Sociological Justice

# 1

# *Introduction*

In 1972, the *Yale Law Journal* published a manifesto called "The Boundaries of Legal Sociology."[1] It contained an indictment of sociologists for confusing facts and values in the study of law—what *is* and what *ought* to be—and argued for a reconstruction of the field as a branch of science: a pure sociology of law. Such a field would address legal reality alone—the facts—and pass over in silence all matters of a critical nature. At the time, many legal scholars regarded this proposal as peculiar if not incomprehensible. Some believed that a purely scientific sociology of law was impossible, while others granted that it might be possible but wondered why such a project would be important or interesting. What could it accomplish? Why bother?

The field known as legal sociology had long been preoccupied with the *effectiveness* of law, a comparison of legal reality to a standard of some kind, whether a statute, constitutional doctrine, judicial decision, or vaguer ideal such as the rule of law, due process, or fairness. These inquiries virtually always discovered a gap between the standard and the reality, the law in theory and the law in action, and so the legal process was continually exposed as ineffective and in need of reform.[2] Such conclusions were presented in the language and style of science and were meant as a scientific critique of law. "The Boundaries of Legal Sociology," however, suggested that a concern with legal effectiveness obscures the difference between science and policy. How law should operate is a question of value, not fact, and since sociology—like any science—can deal only with facts, it cannot assess the effectiveness of law or anything else. A scientific critique of law is illogical and impossible, a contradiction in terms. The reigning version of legal sociology was therefore doomed to failure. But another approach seemed feasible.

The sociology of law could be truly scientific in spirit and method, unconcerned with policy and uncontaminated by practical considerations.[3] Law could be studied as a natural phenomenon. The goal could be a general theory capable of predicting and explaining legal behavior of every kind. This could be done for its own sake, and nothing more. Pure science.

Now all of this has changed. A scientific version of legal sociology has emerged and flourished, and this has happened more rapidly and dramatically than anyone could have imagined in the early 1970s. The field has a new paradigm and a growing inventory of findings and formulations. It is spreading into universities and research centers throughout the world, even into American law schools, sanctuaries of conventionality in legal scholarship.

This book explores the significance of sociological knowledge about law for law itself. As this knowledge expands, legal life will inevitably change. Legal sociology has applications in the practice of law, in the reform of the legal process, and in jurisprudence and social policy. Law is entering an age of sociology.

## The Sociology of the Case

Much of legal sociology focuses on a subject that has long preoccupied lawyers: cases. Only this part of the field—the microsociology of law—will be featured here, since it alone has the peculiar practical and jurisprudential implications later chapters will address. The sociology of the case may be contrasted with the macrosociology of law, a broader inquiry into how legal doctrines and institutions reflect the society and culture in which they occur. The classical era of the field—before the study of legal effectiveness—was largely macrosociological in approach,[4] and this tradition still has many adherents, especially among those influenced by Karl Marx.[5] The earlier approach also seems to be responsible for the popular conception of the field: How does law express cultural values and public opinion? How is it related to economic interests? Questions of this kind might seem obvious and important for any social scientist involved in the study of law, and they are. But only the sociology of the case is new.

Although the sociology of the case as a self-conscious field of research and theory is recent, its ancestry can be traced to earlier scholarship. This includes most notably the movement known as legal realism that began among turn-of-the-century American lawyers and continued for several decades.[6] The central claim of legal realism was that the doctrines of law—the rules and principles—do not by themselves adequately predict and explain how cases are decided. Judges and juries typically decide cases according to their personal beliefs and feelings, and only afterward turn to the written law for a justification. Thus, a famous adage of legal realism was that judicial decisions often have less to do with legal precedent than with what the judge had for breakfast. Although the legal realists engaged in little research and formulated little theory about the reality of law (the above "digestive theory of law" notwithstanding), they challenged conventional thinking about legal decisionmaking and took the first step that modern sociologists would later follow.[7]

A second development in the genealogy of the field has been the gradual accumulation of facts about how cases are actually handled in everyday life. Strangely enough, research on this subject began not in the modern courtrooms that inspired the legal realists, but rather in settings more familiar to explorers, missionaries, and anthropologists. For example, the first systematic study was based on the recollections of Cheyenne Indians interviewed by an anthropologist in collaboration with a lawyer who was prominent in the legal realism movement.[8] The handling of cases was first directly observed for scientific purposes by an anthropologist among the Barotse tribe in what is now the African nation of Zambia.[9] Numerous studies followed, not only in tribal and peasant societies throughout the world,[10] but also increasingly in modern settings, especially the United States. Much of the research in modern America has been concerned with criminal justice, including police work,[11] plea bargaining between prosecutors and defense attorneys,[12] and court dispositions.[13] The handling of civil and regulatory cases has received attention as well.[14] Topics studied include the conditions under which residents of a government housing project resort to law,[15] those under which affluent suburbanites do so,[16] the handling of personal injuries in a small town,[17] the handling of homicides in a large city,[18] how urban dwellers express grievances against their neighbors,[19] how cattle ranchers do so,[20] and the handling of complaints by consumers in the marketplace.[21] Investigations in other societies have focused on such diverse topics as police work in Britain,[22] Holland,[23] and France,[24] dispute settlement in a Swedish fishing village[25] and in Bavaria,[26] Turkey,[27] and Lebanon,[28] popular tribunals in Cuba,[29] and criminal and civil justice in Japan.[30] Historical work has been rapidly accumulating as well, such as studies of legal life in Republican Rome,[31] Aztec Mexico,[32] fourteenth-century Venice,[33] sixteenth- and seventeenth-century Spain,[34] Manchu China,[35] early England,[36] and colonial and post-colonial America.[37] In short, knowledge about the handling of legal cases in societies throughout the world and across history has grown enormously. And we can now corroborate what the legal realists had always claimed: Generally speaking, legal doctrine alone cannot adequately predict or explain how cases are handled.

Countless studies such as those cited above demonstrate that technically identical cases—pertaining to the same issues and supported by the same evidence—are often handled differently. People may or may not call the police or a lawyer; if they do, a prosecution or lawsuit may or may not result; some defendants lose while others win; the sentence or civil remedy imposed changes from one case to another; some losers appeal and others do not; and so on. In other words, law is variable. It differs from one case to the next. It is situational. It is relative.

Consider, for example, murder. In modern America, the official reaction to cases that technically qualify as criminal homicide ranges along a continuum from almost no response at all to capital punishment. If the police find the body of a skid-row vagrant and the circumstances indicate that he died violently, such as by a beating or stabbing, the routine classification of the case in at least one American city is "death by misadventure."[38] It is not handled as a crime. The matter is closed without further investigation and is not even recorded in the official crime rate. Sociologically speaking, the intentional and malicious killing of a skid-row vagrant is not a crime at all. In another city, grand juries refuse to indict over one-third of those arrested for killing a relative, friend, or other close associate, despite what seems to be overwhelmingly incriminating evidence.[39] Trial juries often refuse to convict homicide defendants under similar conditions, particularly when sexual jealousy is involved.[40] At the other extreme are those receiving capital punishment, and in between are all the other possibilities: those who plead guilty to a lesser charge, those convicted but released on probation, those sentenced to a brief period in prison, those sentenced to five years, ten years, twenty years, or life. This variation occurs among and between cases that the written law does not distinguish.

We can also observe differences of this sort in the handling of other crimes, in civil matters such as negligence and breach of contract cases, and in regulatory matters such as consumer- and environmental-protection cases. Citizens do not even notify the police or a lawyer in the vast majority of cases in which they have a legally relevant complaint, and, if they do, formal action by a police officer or lawyer is highly unlikely anyway.[41] In civil matters involving $1,000 or more, for example, individuals in the United States contact a lawyer in only about one-tenth of the cases that technically qualify for a trial in court.[42] When contacted, lawyers file a formal complaint in only about half of these cases.[43] If a lawyer files a complaint, the parties agree to an out-of-court settlement in over 90 percent of the cases.[44] Hence, of all the civil cases involving at least $1,000, including those where no lawyer is contacted and those where no formal complaint is filed, fewer than one percent result in a courtroom trial.[45] Finally, if a trial does occur, the range of outcomes varies from acquittal and token restitution to life-destroying punishment and millions of dollars in damages. Again, legal doctrine cannot adequately predict or explain any of this variation. What, then, does?

In addition to technical characteristics—how the doctrines of law apply to the facts—every case has social characteristics: Who has a complaint against whom? Who handles it? Who else is involved? Each case has at least two adversaries (a complainant or victim and defendant), and it may also include supporters on one or both sides (such as lawyers and friendly witnesses) and a third party (such as a judge or jury). The social characteristics of these people constitute *the social structure of the case*. What is the social standing of each? How much social distance separates them? Each might be higher or lower in social status, for example, and more or less intimate with the others. The adversaries might be relatively wealthy or poor, or one might be wealthier than the other but less conventional or respectable. One might be an organization and the other an individual, one might be more or less integrated into the community, more or less educated, and closer or further from the other in ethnicity, religion, or lifestyle. The status of each supporter similarly contributes to the social structure of the case, as does the social distance between each supporter and everyone else. What, for instance, are the characteristics of the lawyers? Were they acquainted before the case arose? If so, how closely? The characteristics of each third party are variable as well. Does the judge represent a national or a local government? What is the judge's race, ethnicity, social background, and financial standing? Did the judge know any of the parties prior to the case at hand? Who are the jurors?

Every case is thus a complex structure of social positions and relationships. In recent years we have accumulated a great deal of evidence showing that this structure is crucial to understanding legal variation from one technically identical case to another. We have discovered that the social structure of a case predicts and explains how it is handled.

## Law as a Quantitative Variable

The social structure of a case is relevant to every kind of legal behavior, including the likelihood of a telephone call to the police, a visit to a lawyer, an arrest, a prosecution, a lawsuit, a victory in court, the severity of a disposition, and the likelihood of a successful appeal. One phrase describes all of this: *variation in the quantity of law*.[46]

The quantity of law is the amount of governmental authority brought to bear on a person or group. Each legal action against a defendant is an increment in the overall quantity of law that a case attracts. For example, an arrest is more law than no arrest, a conviction or other decision against a defendant is more law than a dismissal, a greater amount of punishment or compensation is more law than a lesser amount, and a successful appeal by a complainant is more law than a successful appeal by a defendant. We now possess a body of sociological theory that predicts and explains variation in the quantity of law in all these instances and more. This theory is still new and undoubtedly will need refinement and elaboration in the years to come. Even so, it indicates how the social structure of a case is relevant.

## Adversary Effects

Who has a complaint against whom? In a modern society such as the United States, the social structure of the complaint itself may be the most important predictor of how a case will be handled. What, for example, is the *social status* of the adversaries?

Social status has a number of dimensions, such as wealth, education, respectability, integration into society (by employment, marriage, parenthood, community service, sociability, etc.), and conventionality (in religion, politics, lifestyle, etc.). Although each kind of social status is a factor in legal behavior,[47] for present purposes we can treat all of them together. In this general sense, social status has long been popularly regarded as a source of variation in legal life—often called discrimination—though this has been disputed by other observers. Some claim, for instance, that poor and black defendants in American courts receive more severe treatment than those who are wealthy and white, but others insist this is not true. What legal sociologists have learned, however, is that both of these popular views are wrong.

A sizable body of evidence from a number of societies and historical periods indicates that, by itself, the social status of a defendant tells us little or nothing about how a case will be handled. Instead, we must consider simultaneously each adversary's social status in relation to the other's. This will show that any advantage associated with high status arises primarily when it entails social superiority over an opposing party, while any disadvantage of low status arises primarily when it entails inferiority. In fact, a high-status defendant accused of an offense against an equally high-status victim is likely to be handled more severely than a low-status defendant accused of an offense against an equally low-status victim. In modern America, for example, a white convicted of killing a white is more likely to receive capital punishment than a black convicted of killing a black.[48] During a five-year period in the 1970s—in Florida, Georgia, Texas, and Ohio—whites convicted of killing a white were about five times more likely to be sentenced to death than blacks convicted of killing a black.[49] Blacks convicted of killing a black were sentenced to death in fewer than one percent of the cases.[50] Moreover, all known legal systems tend to be relatively lenient when people of low status victimize their peers.[51] In other words: *Law varies directly with social status.*[52]

But when people offend a social superior or inferior, another pattern becomes evident. Those accused of offending someone *above* them in social status are likely to be handled more severely than those accused of offending someone *below* them. Those victimizing a social superior inhabit a legal space all their own, with a risk of severity vastly greater than anyone's. Thus, when a black is convicted of killing a white in the United States, the risk of capital punishment leaps far beyond every other racial combination. In Ohio, it is nearly 15 times higher than when a black is convicted of killing a fellow black; in Georgia, over 30 times higher; in Florida, nearly 40 times higher; in Texas, nearly 90 times higher.[53] Similarly, an experiment in jury behavior found that sentencing for negligent homicide with an automobile is by far the most severe when the victim's status is above the offender's.[54]

On the other hand, crimes against a social inferior receive unusually lenient treatment. When a white is convicted of killing a black, for example, the risk of capital punishment is approximately zero. In Ohio, none of the 47 whites convicted of killing a black received a capital sentence during the five years studied; in Georgia, two of 71; in Florida, none of 80; in Texas, one of 143.[55] The most striking comparison is thus between the extremely high likelihood of *downward* capital punishment (where the offender's social status is below the victim's) and the extremely low likelihood of *upward* capital punishment (where the offender's social status is above the victim's). This, too, illustrates a principle of legal behavior throughout the world and across history: *Downward law is greater than upward law.*[56]

We can also rank various status structures from those attracting the most law to those attracting the least. Downward cases attract the most law, next cases between people of high status, then those between people of low status, and upward cases attract the least. Figure 1 depicts these four possibilities. Accordingly, capital punishment in the United States is most likely when a black kills a white, next when a white kills another white, then when a black kills another black, and it is least likely when a white kills a black.[57] This pattern seems to occur in cases of all kinds and at every stage of the legal process.[58]

Now let us turn to a second illustration of how the social structure of a case is relevant. Consider the degree of intimacy between the complainant (or victim) and the defendant: Are they members of the same family, friends, fellow employees, neighbors, or complete strangers? We speak of this as the *relational distance*[59] between the parties. It is one of the most powerful predictors of legal behavior yet discovered.

The closer people's relationships are, the less law enters into their affairs. A grievance between people such as relatives or old friends is likely to result in less law than the same grievance between casual acquaintances or strangers. If, technically speaking, an "assault" such as a beating occurs, for instance, a victim who is intimate with the assailant is less likely to call the police.[60]

Figure 1. Status structure and relative attractiveness to law
(1 = most; 4 = least)

Defendant's Status

|  |  | Low | High |
|---|---|---|---|
|  | High | 1 | 2 |
| Complainant's Status |  |  |  |
|  | Low | 3 | 4 |

If they handle a case involving intimates, the police are less likely to regard it as a crime[61] or to make an arrest.[62] If an arrest is made, the prosecutor will be less likely to bring formal charges;[63] and if the case does go to court, a conviction and prison sentence will be less likely.[64] Guilty pleas—self-applications of law—obey the same principle: People who victimize their intimates are less likely to cooperate with the authorities and plead guilty.[65] The frequent use of courts by married people seeking a divorce may seem an exception to the tendency of intimacy to retard the application of law, but this occurs only where the state certifies marital relationships and prohibits their dissolution without its authorization. Where the state is not a party, people simply end conjugal relationships without resorting to law, a practice seen in numerous tribal societies[66] as well as among formally unmarried couples who live together in modern societies.

The principle of relational distance applies to the handling of homicide, rape, theft, and other criminal matters.[67] In the United States during the late 1970s, for example, those convicted of killing a stranger were considerably more likely to receive a sentence of capital punishment than those convicted of killing a relative, friend, or acquaintance. Stranger killers were four times more likely to receive the death sentence in Florida, six times more likely in Illinois, and ten times more likely in Georgia.[68] The same principle applies to civil matters: People are less likely to litigate a breach of contract that occurs in a longstanding business relationship,[69] and friends or neighbors are less likely than strangers to litigate a case of negligence.[70] Contract or negligence litigation between members of the same family is almost unthinkable and in some cases legally impossible. And again the pattern occurs across societies and history.[71] Intimacy everywhere provides a measure of immunity from law. If we know that the relational distance between the adversaries differs in two otherwise identical cases, we can predict which is likely to attract more law (a formal complaint, a victory for the prosecution or plaintiff, a more severe outcome for the defendant, etc.). We cannot be certain, but our ability to anticipate the result is vastly improved. In sum: *Law varies directly with relational distance.*[72]

The status of the adversaries and their degree of intimacy by no means exhaust the social factors associated with legal variation. Also relevant are the cultural distance between the adversaries, their degree of interdependence, whether they are individuals or organizations, the extent to which alternatives to law are available, and other variables.[73] More details will be presented later. But we should note here that the social characteristics of the adversaries predict the handling of cases only if those characteristics are known. The social status of a complainant or defendant, for example, is relevant in court only if the judge or jury learns about it. The amount of this information about the parties is thus an important factor in itself: The more there is, the more social differentials in the handling of cases are possible (see Chapter 4).[74]

## Lawyer Effects

As mentioned earlier, the social structure of a case is defined not only by who has a complaint against whom, but also by who supports whom and by who intervenes as a third party. Supporters such as lawyers, witnesses, and interested bystanders openly take sides, and their social characteristics have the same pattern of impact as those of the adversaries. The degree of impact, however, depends on the participation of each. Because lawyers have a greater role in the proceedings, their social characteristics are more consequential than those of the witnesses, for example, but we must take into account the characteristics of all the supporters together when calculating the legal chances of each side. For purposes of illustration, consider the lawyers.

Like the adversaries, a lawyer may be higher or lower in social status.[75] Some are wealthier than others, more conventional ethnically or otherwise, more integrated into the community, more respectable, and so on. The legal impact of these characteristics depends on those of the client. Although representation by a lawyer increases the chances of success for anyone involved in litigation,[76] people of lower status, such as poor blacks, benefit proportionately more than those of higher status, such as wealthy corporations, who have advantages in their own right. The higher the lawyer's status, the more benefit the client enjoys. By raising the social stature of lower-status adversaries, lawyers homogenize and equalize the treatment of cases, though they cannot altogether nullify the disadvantages of people opposed by social superiors.

Lawyers may also significantly alter the relational structure of cases. Their impact depends on who the adversaries are—their degree of intimacy with each other—and on the lawyers' ties to each other. If the adversaries are complete strangers, their lawyers may nonetheless be acquainted from previous cases or elsewhere. Even when the lawyers are unacquainted at the outset, they inevitably become closer during the handling of a case merely by communicating with each other about the conflict. And simply because both are lawyers, they have a common bond. Generally, then, lawyers tend to narrow the social distance between strangers and others socially alien to each other when a conflict erupts. This reduces the tendency of socially distant people to use law to its fullest extent and increases the likelihood of a negotiated settlement.

Where the adversaries are socially close, however, the impact of lawyers can be just the opposite. The entrance of lawyers into a conflict between family members (such as a married couple), between members of the same organization (such as fellow employees), or between people in a continuing business relationship (such as a supplier and a long-time customer) often increases the social distance between the two sides, and this in itself increases the likelihood of litigation. The intervention of lawyers may also undermine the ability of the parties in an on-going relationship to resolve their conflict without recourse to law. Once lawyers become involved, a husband or wife may lose influence over a spouse and a supervisor may lose influence over a subordinate, for example, and this too increases the likelihood of litigation. Lawyers may even actively escalate a conflict between intimates. For instance, although one of the spouses opposes a divorce in about one-fourth of the divorce cases handled by lawyers in Holland, the lawyers nearly always argue that "resistance is useless" and successfully pressure these individuals to end their marriage.[77] This suggests that without the participation of lawyers some divorces would be avoided. In short, while lawyers reduce the likelihood of litigation in many cases, in others their effect is precisely the reverse.

*Third-Party Effects*

Who is the judge? The prosecutor? The police officer? The jurors? Are they men or women, old or young, married or single, wealthy or not, white or black, of English, Irish, Jewish, or Hispanic ancestry? Did the judge know any of the litigants or lawyers before the case arose? Are the judge, litigants, and lawyers members of the same community? Here is another component of the social structure of a case: Although sociological research has focused more on the adversaries than on third parties (or supporters), who handles a case also influences how it will be handled.

The behavior of third parties such as judges and jurors varies in its authoritativeness.[78] Authoritative judges and jurors are more likely to select a single winner than to compromise and give something to each side. They are more legalistic—more likely to apply the letter of the law regardless of the consequences—and more punitive and coercive. They generally apply more law when deciding against a defendant, but sometimes decide in favor of the defendant and apply none at all. Less authoritative judges are more lenient and less likely to decide totally in favor of either side.

The degree of authoritativeness varies with the social characteristics of the third party. For example: *Authoritativeness is a direct function of the third party's relative status.*[79] The greater the social elevation above the adversaries and their supporters, the more authoritatively the third party is likely to behave. In modern America, we would therefore expect white Anglo-Saxon Protestant judges to be more authoritative than black or Hispanic or Jewish judges.

And apparently they are: In one northeastern city, for example, white judges are more likely than black judges to convict defendants of serious crimes (felonies). White judges handling black defendants have the highest conviction rate, and black judges handling white defendants have the lowest conviction rate.[80] A comparison of sentencing in two cities indicates that judges of northern European ancestry and those from a middle-class background are more legalistic and severe than ethnic judges (such as Jews and Italian-Americans) and those from a lower-income background.[81] The social characteristics of appellate judges are important as well: Protestant judges of British ancestry are more likely than Catholic judges of non-British ancestry to uphold a conviction.[82]

Since jurors usually are socially inferior to judges and closer to the status of defendants, they are likely to be less authoritative than judges. In fact, juries tend to avoid all-or-nothing decisions in favor of compromises and are less legalistic, and so many lawyers regard them as "unpredictable" and "incompetent."[83] They are also more lenient than judges.[84] And jurors of lower status are less authoritative than those of higher status. A recent experiment showed, for example, that black jurors are less willing to find a criminal defendant guilty than are white jurors.[85] Another discovered that jurors of lower status—such as those with less education and income—are less legalistic, indicated by their lack of explicit concern with the meaning and relevance of legal doctrines during their deliberations.[86]

Jurors of higher status are not only more authoritative but also talk more and have a disproportionate influence on jury deliberations.[87] This means that juries would be even less authoritative if their decisions were not biased by their socially advantaged members. Bear in mind, however, that authoritativeness does not arise from social status in absolute terms but rather from the status of third parties *in relation to* that of the adversaries. While white jurors find defendants guilty more often than black jurors, for instance, they are even more likely to do so when the defendant is black.[88]

Third-party behavior varies with its relational structure as well. Since intimacy breeds partisanship,[89] either side of a case would typically benefit from having a relative or other close associate as the third party. But this is rarely possible. Judges, jurors, and other third parties normally disqualify themselves whenever such intimacy is present; failing to do so, they can and often will be challenged by the other side. Third parties therefore tend to be equally intimate with the adversaries.[90] But this applies only to extreme forms of closeness such as family ties, friendships, or business and professional partnerships. Other disparities are common, such as where one party is an out-of-towner while the other is not, or one lawyer is well known to the judge while the other is not. A recent study indicates that in a small-town court a party from out-of-town suing someone well known to the jury has almost no chance of winning.[91] Being a stranger is probably as great a disability as being, say, poor and black. Even the smallest degree of intimacy, such as eye contact with jurors, strengthens a case.[92]

Unlike judges and jurors, police officers are not subject to challenge when they are close to one of the parties in a case. Hence, police work illustrates what can happen when the social distance between a third party and the adversaries is highly skewed. For instance, a citizen bringing a criminal complaint against a police officer typically finds that the officer's colleagues side with their colleague from the beginning and rarely even pretend to be impartial.[93]

Most third parties, however, are socially equidistant from the adversaries. The third party typically stands at the apex of an isosceles triangle of intimacy.[94] And the dimensions of this tri-

Figure 2. Relational structure and authoritativeness of third party

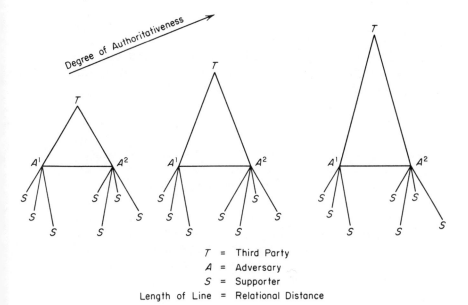

T = Third Party
A = Adversary
S = Supporter
Length of Line = Relational Distance

angle predict how the case will be handled: *The authoritativeness of the third party is a direct function of the relational distance between the third party and the adversaries.*[95] The farther away the third party from the adversaries—the greater the height of the triangle of intimacy—the more authoritative the settlement is likely to be. For this reason, the handling of similar cases varies across societies with differing amounts of social distance between third parties and adversaries. In simple tribes where the third parties know everyone, for instance, decisions are less authoritative than in modern societies where the third parties are often complete strangers.[96] Figure 2 portrays this pattern. Accordingly, a judge who is a complete stranger to all the parties in a case is likely to be more legalistic and decisive than a judge in a small town who knows the parties. Judges and juries relationally close to the adversaries more often find merit on each side and search for a compromise. Relational distance thus has the same effect as the social elevation of the third party. So does cultural distance, such as between ethnicities, occupations, genders, and generations.[97] The more socially removed a judge or jury, the more decisive the result.

## A Note on Speech

It should finally be mentioned that the success of litigants also depends on how they speak. Recent experiments show that the credibility of people in court increases if they testify in a style characteristic of high-status people.[98]

We can distinguish between "powerful" and "powerless" speech by witnesses in courtrooms. Powerful speech involves fewer hedges such as "sort of" and "kind of," for example, fewer fillers such as "uh," "um," and "let's see," fewer questions directed at the examiner by the witness, fewer deferential expressions such as "sir," and fewer intensifiers such as "very" and "surely."[99] When the content of what they say is exactly the same, those testifying in the powerful mode have more credibility. To a judge or jury, they seem more competent and trustworthy.[100] Since people of higher status are more likely to speak in this fashion, witnesses such as professionals and executives generally have more credibility than manual laborers or clerks, whites more than blacks, men more than women, and so on.

Another finding from these experiments is that people who testify in a narrative or verbose style have more credibility than those who give only short answers to direct questions.[101] This too is associated with social status, perhaps because those of higher status presume the right to talk at will and at length, and because lawyers more readily yield control of the interchange to such individuals. Still another finding is that those who interrupt the lawyer examining them and who struggle for control of the interchange—seemingly more common among people of higher status—have the same advantage: They are more persuasive than witnesses who are docile and deferential.[102] Speech might also reveal foreign or regional origins. In various ways, then, how people speak allows the social structure of a case to insinuate itself into the courtroom when it might otherwise be unknown.

## A New Model of Law

What has emerged in recent years is not merely a new version of legal sociology, but a new conception of law itself. The sociological model of law differs radically from the jurisprudential—or lawyers'—model that has long dominated legal thinking in the Western world.

In the traditional conception, law is fundamentally an affair of rules.[103] The explanation of a legal decision normally lies with one or more rules by which the established facts are assessed. By contrast, the sociological model directs our focus to the social structure of a case—to who is involved in it—and this explains how it is handled. The rules provide the language of law, but the social structure of the case provides the grammar by which this language is expressed. In the jurisprudential model, the social structure of the case has no relevance at all (unless it is mentioned in the rules themselves, such as when the written law exempts a husband who rapes his wife from criminal prosecution). Each case is analyzed in a social vacuum. It is even considered improper and a violation of law itself if the social characteristics of the parties influence the handling of a case (again, unless the rules themselves mention those characteristics).

The jurisprudential model regards law as a logical process. The facts of each case are assessed in light of the applicable rules, and logic determines the result. If the rules define a killing with "malice aforethought" as murder, for example, and evidence in the case at hand indicates such a state of mind was present, then it is a murder. But in the sociological model, law is not assumed to be logical. Law is how people actually behave, and that is all. Murder is only what is so regarded in fact, and this may depend on who kills whom rather than logic.

The jurisprudential model also assumes law is constant from one case to another. It is written and available to all, and the same facts result in the same decisions. In other words, law is universal, applying to all cases in the same fashion. The sociological model, however, assumes law is variable. It changes from case to case with the social characteristics of the parties. All that is constant and universal are the principles that predict and explain the results, and these principles are sociological.

The two models are not simply different versions of legal reality with different assumptions about how law works. They arise from different perspectives and have different purposes and goals. The jurisprudential model is a participant's view, employed by practicing lawyers who seek to show how the rules of law lead logically to a particular decision. Judges routinely employ the jurisprudential model as well, since they frame their decisions in terms of how the rules logically apply to the facts. Neither lawyers nor judges explicitly treat the social structure of a case as relevant to the proceedings at hand, and they may not even mention the social characteristics of anyone involved. Contrariwise, the sociological model entails an observer's perspective. It draws attention to the social characteristics of the participants while ignoring the matters emphasized by lawyers and judges. But this is not because any sociologist regards the rules as entirely irrelevant. Every case has a technical core—the rules in the face of the evidence—that can be meaningfully analyzed in the jurisprudential tradition. And, other things being equal—including the social characteristics of all concerned—this technical core is important in the handling of cases.[104]

Figure 3. Two models of law

| | JURISPRUDENTIAL MODEL | SOCIOLOGICAL MODEL |
|---|---|---|
| Focus | Rules | Social Structure |
| Process | Logic | Behavior |
| Scope | Universal | Variable |
| Perspective | Participant | Observer |
| Purpose | Practical | Scientific |
| Goal | Decision | Explanation |

The jurisprudential model is practical, concerned with how cases should to be decided. The sociological model is scientific, concerned with how cases are actually decided. The jurisprudential model is used to reach decisions. The sociological model is used to reach explanations. Figure 3 summarizes these differences.

We may contrast the two models of law more concretely by considering how each regards social differentials in the handling of cases, commonly known as "discrimination." The jurisprudential model—that now prevails in the modern world—regards discrimination as exceptional, a deviation from normality that can and should be corrected. Ideally, the rules alone determine how a case is decided, and those rules generally do not mention the social characteristics of the parties. The sociological model, however, assumes that the handling of a case always reflects the social characteristics of those involved in it. This applies whenever and wherever law is found. It is not merely a matter of differentials according to the race of the parties, their social class, gender, or other characteristics that nowadays attract particular attention. Many other kinds of discrimination exist as well, such as differentials according to the degree of intimacy between the parties, the cultural distance between them, and their degree of organization, interdependence, integration, and respectability.[105]

No one has ever observed a legal system without social differentials. Discrimination is ubiquitous. It is an aspect of the natural behavior of law, as natural as the flying of birds or the swimming of fish. Discrimination is so characteristic of legal life that the concept of discrimination itself is superfluous in legal sociology. It is not problematic. It is taken for granted. Indeed, whether known as discrimination or not, to a large extent it is the subject matter of the field. But although regarded as commonplace in sociology, social differentials have yet to be acknowledged by the legal profession. The jurisprudential model still dominates the thinking and writing of legal practitioners and educators, and they still regard discrimination as exceptional. This will surely change, however, if only because lawyers will eventually discover how legal sociology bears on their work. As the next chapter seeks to demonstrate, this field can increase the effectiveness of lawyers, prosecutors, or anyone else who ventures into the world of law.

# 2

# *Sociological Litigation*

Law students are taught to assess the merits of a case in the light of legal doctrine. They learn the law of homicide, theft, and other crimes, the law of arrest, search, and other aspects of criminal procedure, the law of torts such as negligence and libel, the law of contracts, property, corporations, civil procedure, and so on. They are given to understand that a thorough knowledge of these doctrines is necessary and sufficient to practice law in the best interests of their clients.

## The Strength of the Case

If a client is accused of murder, for example, the evidence must be assessed to determine the "strength" of the state's case, that is, the likelihood that the prosecutor will win a conviction. Law schools teach that the strength of such a case depends entirely on whether the prosecutor has admissible evidence showing that the elements of murder were present—intent, premeditation, and malice—and that the defendant was responsible for it. The defense attorney's assessment of the strength of the state's case then becomes the primary basis on which the client is advised and represented. For instance, the defense attorney must advise a client charged with murder whether to plead guilty (presumably in exchange for leniency) or to plead not guilty and face a trial.

If a plea of guilty is bargained for leniency, a reasonable sentence must be calculated and recommended. If a plea of not guilty is entered, a defense must be planned that will most effectively expose the weaknesses of the state's case in the light of legal doctrine, including judicial decisions of the past (precedents). In fact, however, if an attorney seeks to pursue a client's best interests, something else must be taken into account: the social structure of the case.

Who allegedly killed whom? That is, what were the social characteristics of the alleged victim and the accused? Was this allegedly an upward or a downward murder (was the accused's social status below or above the victim's)? Or was it a lateral murder (between equals)? If lateral, at what status level were the principals: low, intermediate, or high? And what was the relational structure of the crime? Were the victim and the accused acquainted? If so, how well? Who are the witnesses for each side? Who are the prosecutor and the judge? What are the defense attorney's own characteristics? As outlined in Chapter 1, these matters are crucial to how a case will be handled. If it goes to trial, they will be relevant to the result. If a plea of guilty is bargained for leniency, they will be relevant to the final disposition. In other words, the strength of the case is a sociological as well as a legal question. Variation in the social structure of cases explains many differentials in legal life that the written law alone cannot explain. Hence, anyone who ventures into the legal world without knowing how to assess the sociological strengths and weaknesses of a case has a disadvantage. Any law school that does not offer a course on this subject is denying its students valuable knowledge about how law actually works.

Many lawyers already know from years of experience that the social characteristics of a case can be important. Legal educators commonly ignore this subject, but legal folklore acknowledges that the rules of law may not predict the outcome of a case and that who has a complaint against whom, in whose court, with what jury, can also be significant. Many practicing lawyers have their own theories about the relevance of social variables, especially the characteristics of jury members. Some believe, for instance, that black jurors are more lenient toward blacks and that male jurors are more lenient toward rapists. Some believe that rich and powerful plaintiffs—such as business corporations—have advantages, but others believe that plaintiffs who engender sympathy are more successful. Lawyers hold conflicting theories about defendants' characteristics as well, some believing that social status is advantageous, others that it may be a liability, especially for organizations sued by individuals. A folk sociology of law has thus evolved among members of the legal profession even if, like folk medicine, it is sometimes inconsistent and contrary to factual evidence. Clearly a potential market exists among lawyers for sociological knowledge applicable to their work.

## Sociology in the Practice of Law

The self-conscious application of sociology to the practice of law might be called *sociological litigation*.[1] The forms this might take in the coming years are only dimly imaginable, but we can speculate about some aspects of legal practice it could involve. These include screening cases for professional attention; scheduling fees; choosing case participants; deciding whether to seek out-of-court settlements; preparing cases for trial; selecting judges, jurors, and venues; devising tactics during a trial; and, when losses occur, deciding whether to appeal to a higher court. For now, we ignore the question of whether sociological litigation would be socially desirable and consider only how it might be done.

### Screening Cases

A sociological lawyer would screen clients and select cases according to their social characteristics. Other things (including technical considerations) being equal, preference would go to plaintiffs with sociologically strong cases and defendants facing sociologically weak cases. For example, downward lawsuits— where the plaintiff-client is socially superior to the defendant— would be preferred to upward or lateral lawsuits, and so would those against strangers or others relatively distant from the client in social space. As explained in Chapter 1, these cases are easier to win. In regard to defendant-clients, the calculus is precisely the reverse: Upward cases and relatively intimate cases would be preferred, since judges and juries are less likely to rule against these defendants. In criminal matters, the social characteristics of the victim, the accused, and the accused-victim relationship would similarly be relevant considerations for a prosecutor or defense attorney. But this is not to say that a sociologically astute lawyer would accept only cases with an advantageous social structure. Winning is not everything, particularly for private attorneys. Even losing lawsuits can be effective weapons in social conflict, and some attorneys may believe that all defendants deserve representation, regardless of their legal or sociological situation. Some might also be interested in fees.

### Fees

When a client seeks a monetary award for damages, an American attorney has a choice of fee arrangements, including payment on an hourly basis, a flat fee, or a fee contingent on the amount of money, if any, awarded in court or provided in an out-of-court settlement. An economically rational attorney would choose among these alternatives partially on the basis of the social structure of the case. If a case is sociologically weak (such as when a client wants to sue his wealthy father), despite the possibility of a large monetary award or settlement the attorney might prefer to be paid on an hourly basis or with a flat fee independent of the result. This might apply even when a sociologically weak case is technically strong. If, however, a case is sociologically as well as technically strong (such as where a wealthy client is suing a less prosperous stranger), the attorney might prefer a contingent fee providing a fixed portion of any award the client might win. Contingent fees could also be adjusted accordingly: The worse the case, sociologically, the greater the portion of potential winnings the attorney might request from the client.

In theory, an attorney might design an entire practice from a sociological standpoint. This would involve, for example, locating the practice in a setting with considerable inequality of wealth, cultural diversity, and social anonymity, and one populous in both people and organizations—all conditions under which litigation flourishes.[2] It might also be profitable (as many lawyers already seem to realize) to direct the practice toward those known to use law most frequently and successfully, such as wealthy individuals and business organizations.[3]

## Designing Cases

Occasionally an attorney is able to choose clients or opponents in a particular case. This may happen when more than one person or organization has the same grievance or appears liable for the same wrongful conduct, such as when a number of people are killed in the same accident, injured by the same defective product, or denied housing by the same landlord. On the other hand, a single individual might have the same complaint—say, discrimination in hiring or housing—against several different employers or landlords. Or a corporation's attorney may be able to choose among several technically similar negligence complaints—such as a number of claims concerning the same defective product—to oppose in court (and which would establish a precedent for the others). These situations present lawyers with an opportunity to design the social structure of their cases.

Many lawyers have their own theories about how such choices should be made. One, which might be called "the theory of deep pockets," holds that it is always preferable to bring lawsuits against individuals or organizations who are relatively wealthy. (Lawyers sometimes refer to wealthy defendants as those with deep pockets.) These opponents can better afford monetary losses and, for this reason alone, judges and juries may be more inclined to decide against them. In addition, some lawyers believe that the most successful plaintiffs are those who engender sympathy, as would the desperately poor, the widowed, the unemployed, and those with many children to feed. Individuals are also believed to engender more sympathy than organizations, especially business organizations. This could be called "the theory of the pathetic plaintiff." Law students commonly hear both of these theories in the corridors if not in the classroom, and many undoubtedly apply them when they begin practicing. Both theories are popular with the general public as well, perhaps because ordinary individuals who win large damage awards from wealthy corporations frequently receive attention from the mass media. As it happens, both theories are largely wrong. They are not merely useless, but are the opposite of what a sociologically trained lawyer would do.

Defendants such as wealthy corporations are obviously better able to pay damages if they lose, and judges and juries might order them to pay more as well. Even so, lawsuits against wealthy defendants—especially those brought by a social inferior—are less likely to succeed in the first place (see Chapter 1). While it is true that individuals who sue organizations are increasingly successful in societies such as modern America,[4] a three-city study of civil courts in the United States shows that they are still considerably less successful than individuals who sue other individuals.[5] If it is better to win a little than nothing at all, it would seem more sensible to sue someone such as an individual whose pockets are not the deepest available (as long as they are not entirely empty).

The findings of legal sociology also indicate that a socially "pathetic" person would be one of the worst plaintiffs to represent. Some of these cases, such as matters involving injured children, surely do have a greater chance of success, but individuals of low status are generally less successful in legal life, particularly when their complaint is directed against a social superior. Individuals in the aggregate are also less successful than organizations. For instance, the three-city study cited above shows that organizations bringing lawsuits are more successful than individuals.[6] Moreover, the most successful litigants of all are organizations suing individuals, and the least successful of all are individuals suing organizations.[7] "Pathetic plaintiffs" suing people with "deep pockets" would seem to be among the riskiest clients a lawyer could represent.

Other characteristics of clients and opponents are relevant as well. Plaintiffs relationally distant from their opponents are more likely to win and so would be preferable as clients, for example, whereas defendants are better off when they are as close to their adversaries as possible (see Chapter 1). Lastly, before we consider other principles of sociological litigation, it should be noted that the choice of friendly witnesses involves the same logic as the choice of clients: Those from the highest social level would be preferred. In addition, potential witnesses socially close to the opposing side would be avoided when the client is a plaintiff or victim, since they would reduce the defendant's vulnerability. When the client is a defendant, witnesses close to the opposing side would be desirable for the same reason.

## Pre-trial Decisions

Sociological knowledge can be applied at every stage of the legal process. In American criminal cases, for instance, a major question for the defense attorney is whether the defendant should plead guilty in return for leniency (known as a plea bargain). An attorney might base this decision partly on the social structure of the alleged crime. If the state's case is sociologically strong, such as an alleged robbery of a prominent businessman by a vagrant—an upward crime crossing a considerable distance in social space—the attorney might want to arrange a plea bargain if possible, since the likelihood of a conviction and severe sentence would be relatively great if a trial were to be conducted. When the state's case is sociologically weak, however, as in an alleged assault between friends or lovers in a black slum, the attorney might risk a trial in hopes of an acquittal or, at worst, a lenient sentence. Prosecutors can also take account of these factors and seek an impressive conviction rate by going to trial only when a case structure favors the state. Other things being equal, a prosecutor might offer little leniency in return for a guilty plea—or refuse to bargain at all—when the state's case is sociologically strong (such as in the hypothetical robbery described above) and might bargain generously to obtain a guilty plea when the state's case is sociologically weak (as in the hypothetical assault above).

In civil cases, the lawyer's central question is whether to risk a trial or seek an out-of-court settlement, and the logic is essentially the same as that in plea bargaining. The plaintiff's lawyer might advantageously try to settle a sociologically weak case, such as a lawsuit by a poor individual against a wealthy corporation, since the odds of winning in court would be relatively low. Where feasible, a sociologically weak lawsuit may be strengthened by recruiting an organization to join the case or by recruiting others with a similar problem to join together in a so-called class action. A class action will always strengthen a case if all the participants are at least equal to the original plaintiff in social status and if they are at least equally distant from the opposing side. On the defendant's side, organizational or other support would always be welcomed as well, but a joint trial with other parties in the same case would not always be advantageous. Where fellow defendants introduce sociologically vulnerable characteristics, such as lower status or greater social distance from the plaintiff or victim, the defendant's case might better be severed and handled separately.

## Case Preparation

Preparing a case for trial, a sociologically trained lawyer would search not only for precedents in past decisions involving technically similar cases but also for socially similar cases. A past case essentially identical in respect to the matter in dispute and the evidence presented might nevertheless have an entirely different social structure. If so, the weight of precedent would be less than in cases that are socially as well as technically similar. Suppose everything about a past case of alleged negligence— a hunting accident, let us say—is largely the same as a new case from the standpoint of legal doctrine, but in the earlier case the injured plaintiff was socially superior and unknown to the defendant who fired the gun, whereas in the present case the plaintiff is socially inferior and closely related to the defendant. Suppose further that earlier the court decided in favor of the plaintiff. Obviously an attorney handling the new case should not rely too confidently on the weight of precedent, since an upward lawsuit between intimates is so much less likely to succeed than the earlier downward case between strangers. Attorneys can thus analyze and evaluate precedents sociologically as well as technically.

An attorney should not regard a case as a precedent unless its social structure resembles the case to be decided. This is a strategy of legal research beyond that presently taught in law schools: Instead of limiting the survey of past decisions to technically similar cases—the usual procedure—one searches for cases similar in their social structure, even when technically they are different. Sociological precedents may predict how a case will be decided as well or better than merely technical precedents. If we know that sons who bring negligence suits against their fathers rarely succeed, for example, such a case would be a poor candidate for compensatory damages, regardless of the particular circumstances of the injury. A prudent lawyer for the plaintiff (the injured son) might seek an out-of-court settlement from the defendant (the allegedly negligent father) and, failing this, drop the case altogether.

The social structure of a case includes supporters and third parties as well as the adversaries. Accordingly, cases can be further distinguished by the social identity of these other parties, whether partisan or nonpartisan. Who testified for whom? Who were the attorneys? Who was the judge? Who served on the jury? Because case reports presently contain little information about social characteristics, distinguishing past decisions sociologically is difficult and requires additional research. If the social structure of cases comes to be more widely recognized as an element in legal behavior, however, this information may be routinely gathered and recorded with the other facts now included in case reports.

Lawyers who cannot distinguish cases sociologically as well as technically have a serious handicap. They must forever work in darkness, never understanding why some precedents are upheld while others are ignored. Many decry the unpredictability of the courts, unaware that the behavior of judges and juries is not nearly as mysterious as it first appears. Court decisions are the greatest mystery to those who would understand them with legal doctrine alone. The incomprehensibility of law results from legal education in its present form. But the day may come when law professors will ask their students to distinguish one case from another—the heart of legal education in America—and the answer will be sociological as well as legal.

## Designing Trials

Lawyers often seek to have cases tried before one judge rather than another and to avoid jurors who might weaken their chances of winning. They believe that judges or jurors with particular characteristics are preferable for particular cases. One judge might be known as "hard" on criminals or "sympathetic" to tenants in housing disputes, for example, women jurors might be considered more severe in rape cases, and black and Jewish jurors might be regarded as "lenient" or "liberal" in matters pertaining to members of minority groups. In modern America, jury selection by attorneys[8] may even be based on community surveys, and mock trials may be conducted to learn how citizens with various characteristics would vote in a specific case. Legal sociology can be applied to the selection of judges and jurors as well.

American lawyers often seek to schedule cases with judges or courts thought to be advantageous, a practice known as "judge shopping" or "forum shopping." This can be done sociologically by striving to schedule each trial in the court of a judge as socially close to the client and as socially distant from the opponent as possible. Closeness between the judge and this lawyer and distance between the judge and the opponent's lawyer would also be desirable. Beyond these considerations, the ideal judge would depend on the sociological and technical merits of the case. If the client's case is strong in both respects, as in a downward lawsuit with powerful evidence against an unrespectable stranger, an authoritative judge might be preferred—to settle the case decisively without a compromise. In this way, a lawyer might parlay an initial advantage into a total victory. The same would apply to a defendant-client opposed by, say, a social inferior in a technically weak case. Since the authoritativeness of third parties increases with their social elevation (see Chapter 1), a judge with as many elements of social status as possible might be sought in strong cases—a wealthy white man from a dominant ethnic and religious group, for example. For a strong case, moreover, a jury would be undesirable, since the social composition of juries renders compromise decisions more likely. Should the defendant request a jury under these circumstances, the opposing lawyer might seek to maximize the average status of the jurors by challenging as many socially disabled people as possible (the poor, the unemployed, members of minority groups, etc.). In the United States, another way to raise the status level of the decision in some cases is to file the complaint in a federal rather than a state or other local court, since federal judges are elevated by the social stature of the national government. A federal court might also be preferred if the client would otherwise be an outsider in the opponent's local court.

The preferences outlined above would be reversed when a client's case is weak, sociologically or legally. An indecisive judge or a jury would become more attractive, since a compromise decision—and a partial loss—might be better than risking a total loss with a more authoritative judge. The lawyer would therefore seek a judge and jury with the lowest possible social stature. A sociologically ideal judge and jury, however, cannot be defined categorically. Rather, this depends in every case on the client's willingness to risk total defeat. No matter how strong their case, some clients might be unwilling to take this risk and so would always prefer an indecisive judge or a jury. On the other hand, some with a weak case might still insist on maximizing their chances of total victory and so would avoid a jury and prefer a decisive judge whenever possible. Each lawyer can only calculate the options for each client as an individual and might even leave the final choice to the clients themselves. Sociology is only a tool, not a theology, and cannot specify the proper degree of risk for anyone.

## Managing Trials

Even if the adversaries, witnesses, judge, and jurors have been sociologically screened and selected as much as possible, a relentless application of the sociological strategy would extend to courtroom presentations as well. Here the sociological lawyer would seek to manage information about the social characteristics of the case in the client's best interests. Oral presentations, including the examination of witnesses, can introduce and emphasize the client's advantageous characteristics and the opponent's characteristics known to be disabilities. The lawyer would draw attention wherever possible to every kind of social superiority the client enjoys, such as a history of steady employment, a high income, substantial savings, investments, ownership of a house or business, a prestigious address, socially prominent friends and associates, a high level of education, a spouse and children, civic contributions such as military service or youth work, organizational memberships, and an unblemished reputation. Conversely, the court would learn in vivid detail—when opportunities arise—everything about the opponent that is damaging, such as unemployment, a history of employment problems, financial troubles, a lack of major possessions such as a house or automobile, a lack of education, a lack of roots in the community, a lack of a family, evidence of social isolation, a criminal record, a dishonorable discharge from the military, and unconventionality of any kind. A lawyer can similarly manage social information to enhance or undermine the credibility of witnesses.

The nature of the relationship between the adversaries should be considered as well: If a client is the defendant, the lawyer might wisely draw attention to any element of intimacy between the client and the plaintiff or victim, since this would operate against an adverse decision and, failing that, against severity in the final disposition of the case. If a client is the plaintiff, or if the attorney represents the state in criminal court, any element of social distance separating the adversaries (including the victim and defendant in criminal cases) would be featured. Even if the adversaries are superficially very close, such as members of the same family, the attorney for the plaintiff or prosecution would advantageously draw attention to anything about the relationship indicating it is weakening or coming apart to counteract any sociological immunity the defendant might otherwise enjoy.

The foregoing does not mean, however, that a sociologically sophisticated lawyer merely adapts to the social structure of a case once the trial begins. Not at all. The lawyer can also adjust the case structure as the trial proceeds. For example, a criminal defendant who is unemployed has a disadvantage, but this can be remedied. Even after a conviction it may be possible to find a job for a defendant before sentencing, and this might mean the difference between jail and probation. Some defense lawyers already use this tactic and, when appropriate, encourage clients to improve their social appearance by going back to school or entering drug treatment or other rehabilitation programs. American defense lawyers routinely alter the social identity of defendants (called "manufacturing facts" by one observer).[9] One lawyer remarked that it is wise to "help the defendant begin his rehabilitation as soon as possible. Get a job, make restitution, get him involved in social service programs, [so that] by the time sentencing comes up, the judge is likely to accede to the plan the defense attorney presents."[10]

Defense lawyers in criminal cases also frequently insist that their clients dress and groom themselves to create an appearance of social stature and respectability. Friendly witnesses, family members, and other supporters might be advised to do the same. Prosecutors might similarly prepare the alleged victim and other witnesses, and so might attorneys prepare clients in civil cases.

Speech can be adjusted as well. As noted in Chapter 1, experimental evidence shows that how people talk significantly affects their credibility in the courtroom. Those who speak in a "powerful" style (with fewer hedges, fillers, questions, and expressions of deference) and those who speak narratively and assertively are better able to convince a judge or jury that they are telling the truth.[11] A client's speech might be monitored accordingly and, where possible, corrected to be as "powerful" as possible. The attorney might also screen the speech of potential witnesses and select those who naturally speak in an advantageous manner, perhaps coaching them to speak even more "powerfully" before they testify in court. Since the credibility of written language obeys the same principles,[12] out-of-court depositions might be planned in the same way.

Once in the courtroom, an attorney can continue to take account of social variables known to be associated with success or failure. For example, hostile witnesses who speak in a fashion that strengthens their credibility might be cross-examined as quickly as possible and with questions allowing only "yes" or "no" answers. Where feasible, it might be better not to cross-examine them at all. But hostile witnesses whose speaking style weakens their credibility might be kept on the stand and required to talk as long as possible. The speech of client-defendants in criminal cases would also be considered before calling them to the witness stand.

The relationship between a client and those deciding the case is even manipulable. Since, for example, any increase in the degree of intimacy between a criminal defendant and the judge or jury enhances the likelihood of a more lenient outcome, every tactic having this effect might be implemented as a standard procedure. Other things being equal, the defendant-client should be kept on the witness stand as long as possible to narrow the social distance separating this individual from everyone else in the courtroom. During the oral examination, rich detail about the defendant's private affairs and biography increases intimacy with those present. Eye contact with the judge and jury might also be recommended to clients, friendly witnesses, and other supporters.[13]

Since plaintiffs and victims of crime increase the strength of their cases in proportion to the strength of their bonds with judges and jurors, the same tactics might be adapted for them. The lawyer, too, might get as close as possible to those deciding a case. In fact, simply prolonging a trial increases the overall level of intimacy in the courtroom. Insofar as this narrows the distance between the adversaries, it benefits the defendant (since law varies directly with relational distance and inversely with intimacy). However, insofar as it narrows the distance between the judge or jury and the adversaries (and their supporters), it tends to reduce the authoritativeness of the court, increasing the likelihood of a compromise decision that gives something to both sides. Many litigants would welcome a compromise, but not all. Those with a very strong case might regard a compromise as a loss. If so, lawyers handling strong cases should seek fast trials.

## Appeals

Although a well-trained lawyer would normally try to avoid going to trial when either the technical or the sociological aspects of a case are unfavorable, trials are sometimes unavoidable,

and every lawyer sometimes loses. When this happens, the decision whether to appeal to a higher court can be made partly on sociological grounds.

First is the question of whether and how much the trial decision contradicted one or more sociological principles. An appellate court might be expected to correct glaring contradictions—sociological error—though not explicitly or even consciously. Holding constant the technical elements, for instance, a client's downward lawsuit against a stranger that is lost at trial is a better candidate for reversal than an upward case against an intimate.

Many cases are neither sociologically nor technically one-sided, but the decision whether to appeal to a higher court can still be calculated sociologically. Since appeals are argued without the active participation of the adversaries or their supporters, the characteristics of the attorneys—including their social characteristics—loom larger at this stage. A particular case may thereby become sociologically stronger or weaker when it is appealed. For that matter, a trial lawyer eager to reverse a loss might enlist a new lawyer, such as a specialist in appellate litigation more socially prominent and closer to the appellate judiciary, for the appeal.

A sociologically oriented lawyer would also consider the social characteristics of the appellate court itself. The social status of judges is almost always higher at the appellate level, if only because of the greater stature of the governmental unit they normally represent (e.g., a state or federal government rather than a local government). Appellate judges are the superiors of trial judges not only in the legal system but frequently in other respects as well, such as in social background and material well-being. These attributes increase the authoritativeness of appellate judges, lending a more legalistic and decisive character to their behavior. This might benefit a defendant appealing a compromise decision by a jury of working-class individuals, but it would offer less hope for someone who lost a bench trial before a socially elite—and already legalistic and decisive—judge. The same calculus applies at each tier of the appellate system, from lowest to highest, so that a client's chances of winning a reversal might rise or fall as the case moves from one level to the next. Hence, the wisdom of appealing a case to "the highest court in the land" is partly a sociological question.

## The Unauthorized Practice of Sociology

The principles of sociological litigation may be useful to laypersons as well as lawyers. For example, a private individual or organization might want to assess the sociological merits of a problem before contacting an attorney. When existing sociological knowledge suggests that success is unlikely, the matter might be handled informally or not at all. This practice, if widespread, could substantially reduce attorneys' earnings, but once legal sociology escapes its ivory tower, applications by nonlawyers will surely begin. Unlike the conventional practice of law, sociological litigation is not likely to be monopolized by a closed profession or guild. Sociologists will probably never demand legislation banning the unauthorized practice of their profession.

Just as lawyers might screen potential clients on sociological grounds, so the reverse can be done as well: Clients-to-be might screen potential lawyers on the basis of *their* social characteristics. An attorney with all the trappings of social status, such as an older white man in a large and prosperous law firm, might then be chosen to enhance the strength of a case, whereas younger attorneys trying to establish their practice—especially women and members of minority groups—might be avoided. And to increase the likelihood of an out-of-court settlement, a client-to-be might search for a lawyer as intimate as possible with the opponent's lawyer. If a trial threatens to occur, a potential lawyer's closeness to the court might enter into consideration. The client-to-be might prefer a local attorney to an outsider and, among locals, one who has practiced for many years and is tightly integrated into the community. But for an appeal, a replacement socially closer to the higher court would be desirable.

Criminals and other lawbreakers could also conceivably take sociological knowledge into account when planning their nefarious activities. Other things being equal, for instance, it would be advantageous for them to select their victims from among the poor, the marginal, the unconventional, the uneducated, the unrespectable, and others lacking in social stature and relatively remote in social space since, when caught, defendants in these cases are less likely to be convicted or otherwise subjected to law. If chosen carefully, many victims would not even complain.

In the coming years, an increase of sociological litigation would seem inevitable. In light of recent discoveries, after all, any lawyer who fails to practice law sociologically is missing an opportunity for greater effectiveness. Arguably, a totally unsociological lawyer is incompetent. Since lawyers are ethically and legally bound to serve their clients as conscientiously and skillfully as possible, those wishing to practice law may someday be required to study legal sociology in law school and apply it when handling cases. Unsociological litigation might even come to be considered a form of malpractice. Nevertheless, many readers will surely find the idea of sociological litigation unattractive.

The application of sociology to litigation might well shock and disgust anyone who believes in the rule of law. That the social structure of cases predicts and explains how they will be handled by judges and juries might seem bad enough, a gross violation of the principle of equal treatment and a form of discrimination. But the prospect of lawyers taking this state of affairs for granted and seeking to exploit it for their own advantage must seem still worse.[14]

Yet indignation alone cannot change these developments. Like everything else in the universe, law varies with its environment. How cases are handled depends on their social context. It is unrealistic to expect lawyers or anyone else knowingly to ignore this brute fact and continue acting as if law were a matter of rules and logic and nothing more. Since law reflects its social environment in predictable patterns, however, by changing this environment we should be able to change its behavior. How might this be accomplished? How might we reduce social differentials in legal life?

Several reforms are explored in the following chapters. One

would redesign the social structure of cases entering the legal system, another the process by which cases are handled, and a third the jurisdiction of law itself. If adopted, these reforms would reduce discrimination to a level never before seen in any legal system, and sociological litigation would become all but impossible.

# 3

# *The Incorporation of Conflict*

Groups are more litigious than individuals. Some tribal societies discourage or even prohibit lone individuals from bringing cases to court at all. The Shavante of central Brazil, for instance, require that every case be brought by a so-called "faction." Only men can belong to these factions, and so women effectively have no legal rights.[1] And some societies handle particular grievances and liabilities collectively that others would handle individually. For example, when homicide is defined as an offense against the victim's family or clan, the killer's family or clan might be held liable to pay compensation or to suffer revenge.[2]

Organization[3] is important in legal behavior everywhere. Like the social status of the adversaries or the intimacy between them (discussed in Chapter 1), the organization of the adversaries predicts how their case will be handled.[4] Unorganized people—individuals on their own—use less law and are more vulnerable to law. Individuals vastly outnumber organizations in modern societies such as the United States,[5] but organizations initiate more legal cases. Most complaints filed in small-claims courts are brought on behalf of organizations, and nearly all are directed against an individual.[6] In addition, public and private organizations initiate over half of the civil cases pertaining to larger claims, and two-thirds of these are directed against an individual.[7] When crimes such as burglaries or robberies occur, organizational victims are more likely than individual victims to

41

notify the police.[8] Local and national governments also initiate a multitude of criminal cases: The police initiate nearly all vice and traffic cases, for example, and every case of any kind that reaches a criminal court is prosecuted by the state. Moreover, nearly all criminal cases are directed against individuals rather than organizations.

Although lone individuals are the favorite targets of organizations that initiate legal proceedings, the reverse does not apply. Far from it. Individuals are generally reluctant to assert their legal rights against organizations. In modern America, they bring twice as many civil lawsuits against fellow individuals as against organizations,[9] and in legal domains where organizations are the primary wrongdoers, such as consumer and environmental matters, aggrieved individuals typically refrain from legal action altogether.[10] They just "lump it."[11]

## The Organizational Advantage

Not merely more litigious than individuals, groups are also more successful. For example, they win more often.[12] As noted in Chapter 2, organizations bringing civil lawsuits in the United States are more likely to win than individuals, especially when the defendant is an individual rather than another organization.[13] In cases of alleged racial discrimination, a complaint brought to a state commission by an organization (such as the National Association for the Advancement of Colored People) is more likely to succeed than a complaint by an individual. An organization is particularly likely to succeed when the alleged discrimination involves an individual offender rather than another organization.[14] And organizations are more successful when they appeal a lost case, civil or criminal, to a higher court.[15]

Criminal courts are more likely to convict individuals who are accused of victimizing an organization.[16] It is also relevant that criminal defendants in general are more likely to lose than civil defendants: For example, American courts convict the defendant in about 95 percent of the criminal cases, whereas the defendant loses in only about half of the civil cases.[17] This is partly explained by the fact that practically every criminal defendant is an individual opposed by an organization (the state), whereas civil defendants are sometimes organizations and are often opposed by a lone individual. The vast majority of criminal defendants do not even resist the state: About 90 percent convict themselves by pleading guilty.[18] In the rare case where an organization is prosecuted in criminal court, the likelihood of a guilty plea is surely far lower. Another contrast appears when an individual wants to bring a lawsuit against a government: In the United States, this is possible only when the government explicitly consents to be liable for a particular class of injuries. Otherwise it is immune. For instance, the United States refuses liability for injuries suffered or inflicted by members of the armed forces on active duty. The liability of individuals, however, is obviously not a matter of choice.

Societies such as modern America have, so to speak, two legal systems: one for individuals, another for organizations. Criminal justice primarily addresses the misconduct of individuals. Only they are routinely subjected to body searches, arrest, interrogation, harassment, confinement, and other intrusions and deprivations. Only they are executed. For that matter, punishment of all kinds is largely reserved for individuals.[19] Thus, the Shavante of Brazil (the tribal society noted earlier where only groups can complain to a tribunal) only punish individuals. But first the individual must lose the support of his group:

> The community as a whole can only take action against an individual if his faction disowns him. . . . Once a man is either disowned or abandoned by his faction, then he is virtually outlawed. Should there be a serious accusation made against him he can only flee to another community, for a decision to execute him inevitably follows.[20]

Modern societies are not so extreme, but organizations often enjoy special advantages that effectively immunize them against punishment. In the United States, for instance, the government frequently offers organizations a uniquely gentle legal disposition known as a consent decree, by which the organization (usually a business firm) promises not to violate a particular law in the future. This entails no admission of guilt or even official accusation of misconduct, though alleged misconduct lies behind the procedure in every case. In fact, the agreement normally allows the organization explicitly to disclaim having engaged in any such misconduct at any time in the past. Individual law violators such as burglars and robbers would surely prefer a disposition of this kind as well, but it is not available to them. Whether as complainants or defendants, then, organizations are better off than individuals.[21]

It is widely recognized that legislation often favors interest groups and organizations such as business corporations and labor unions. The point here is that a similar pattern occurs in the handling of legal cases. One of the most extreme forms of social bias in modern law, it could be regarded as organizational discrimination.

## Legal Individualism in Modern Society

Despite the advantages of organization, legal conflict in contemporary societies remains highly individualized. This situation has evolved over the centuries, owing primarily to the gradual disintegration of the traditional family with its extended ties between each individual and numerous others, all of whom—especially blood relatives—historically stood together in times of conflict.[22] In many societies, for example, the clan would pay compensation on behalf of a member who victimized a nonmember.[23] But collective liability of this sort has all but disappeared. The breakdown of the traditional family has left individuals in conflict to fend for themselves, without primordial allies, able only to buy help on the open market from lawyers and insurance companies.[24] Perhaps this partly explains the corresponding growth of criminal justice: Under modern conditions, who can pay compensation or otherwise negotiate a settlement for offenders such as common burglars and robbers? Typically they are lone individuals, too poor to make amends themselves and without allies willing or able to do so. No one is available to replace the families who once contributed their own wealth to satisfy injured parties and their kinfolk. Punishment thus becomes the only way for common criminals to "pay" their "debt." Wrongdoers are generally all the worse off if their adversary is an organization, which it frequently is. Then they are more likely to be prosecuted or sued, to lose, and to lose more of whatever is at stake. Still worse off are people in conflict with the state. Like those who get involved in tribal conflicts without a family—such as bastards, foreigners, and other isolates[25]—they are almost certain to lose.

A lack of organization is one of the greatest disadvantages—possibly *the* greatest—anyone can experience in legal life. Nevertheless, in modern societies such as the United States, most people with legal problems are completely alone and unorganized, like bastards in a tribe. This is *legal individualism*.[26]

Few if any societies have ever distributed organization as unequally as those of the modern world. Compare again the pattern in tribal societies, where conflict is frequently incorporated by groups—usually families but sometimes villages, castes, age-grades, or other alliances.[27] This is *legal corporatism*. As societies evolve toward the modern form, however, communal groups such as the extended family lose their dominance in everyday life, not only in the management of conflict but in nearly all arenas of human activity. In their place arises a vast population of specialized organizations to perform functions once in their hands, such as the production and distribution of goods and services, the training of the young, the care of the sick and injured, and relief for the poor and homeless. The most fateful transactions in modern life increasingly occur between individuals and formally constituted organizations.[28] And in nearly all these transactions, the organizations have the advantage. They commonly decide what prices will be paid, jobs performed, services rendered, goods produced, wages provided, standards followed, and so on, and individuals must accept the situation. Conflict between individuals and organizations is increasing as well, and the advantage of organizations is again conspicuous.

Another feature of the proliferation of organizations is the growing tendency of individuals to band together formally to deal more effectively with the organizations in their environment. Individuals counter organization with organization of their own. Specialized groups arise to represent otherwise atomized individuals such as employees, consumers, professionals, farmers, students, ethnic and racial minorities, women, and numerous others. Organization begets organization. But in legal life the pattern is different. Lone individuals continue to find themselves in conflict with organizations, and the organizations continue to be particularly successful under these circumstances. Here, for unknown reasons, the tendency of individuals to counter organization with organization has not materialized.[29]

The participation of lone individuals makes possible many of the social differentials—known as discrimination—that characterize modern law. On the one hand, wherever individuals face a corporate adversary, whether a public or a private organization in a civil or a criminal case, they have a disadvantage simply because they are unorganized ("organizational discrimination"). On the other hand, wherever they face any adversary on their own, whether an organization or not, many individuals have disabilities of other kinds. These include, for example, minorities, the poor, the unemployed, the unconventional, and the unrespectable (see the overview in Chapter 1). Such people may join together and form organizations to represent them—ethnic associations, civil-rights groups, labor unions, etc.—but often they do not. And without organizational help in legal conflicts, they are less likely to get what they want. Many individuals are thus doubly disadvantaged in legal life: first, because they are lone individuals who may face organized adversaries and, second, because they are socially disabled, regardless of whom they face. In both respects, legal individualism perpetuates discrimination. One strategy to reduce discrimination in legal life would therefore be to reduce the participation of lone individuals. This would bring law closer to the evolutionary stage already reached in many arenas of modern life, where the introduction of formal organization has revived some of the conditions of tribal life. The organization of individual litigants would dramatically alter the balance of advantage in legal affairs. How, then, might this be accomplished? How might legal corporatism be restored in modern life?

With this problem in mind, let us examine the handling of conflict in a traditional setting with a system of legal corporatism so advanced that it might conceivably provide a model for contemporary societies such as the United States.

## Legal Corporatism in Traditional Society

Most traditional societies rely on existing corporate structures such as families and clans to pursue grievances and right the wrongs of their individual members, but the nomads of northern Somalia—on the Horn of Africa—have taken this practice a step further. They have created a species of group life specifically designed to manage conflict: the *dia-paying group*, which in Arabic means bloodwealth-paying or compensation-paying group.[30] Virtually all adult men belong to such a group, each ranging in size from a few hundred to a few thousand members. Membership is usually based on family lineage and a contractual agreement between each group and each individual wishing to be a member. In 1957, there were 360 dia-paying groups representing six lineages in northern Somalia.

A dia-paying group demands compensation when a member suffers injury from a nonmember, provides compensation when a member injures a nonmember, and settles disputes between its own members. The membership agreement specifies in advance the precise nature of compensatory payments, including the amount to be paid and demanded for each death, wound, or insult[31] suffered or inflicted. The agreement also specifies the proportion of the compensation to be provided or received by the membership at large and by the individual and family directly involved in a particular case. The payment schedule assures that those suffering an injury will receive the largest share of the compensation received from another dia-paying group and that those responsible for an injury will contribute the largest share of any compensation paid to another group. One-third of a payment is typically received or paid by the immediate kin of the individual injured or responsible for an injury, while the remaining two-thirds is distributed among or contributed by the members at large. Generally a man's life is worth 100 camels (a woman's is worth 50 camels). Hence, if a man is killed by a member of another group, his immediate kin will expect to receive 33 camels, while the remaining 67 are distributed among the larger membership of his dia-paying group. If a man kills someone belonging to another dia-paying group, the killer and his immediate kin must provide 33 camels, while the other members of his dia-paying group provide the remaining 67.

Dia-paying agreements specify how compensation will be dis-tributed or collected among the members at large as well. Each family is given or assessed a particular share of the total. Some dia-paying groups calculate the precise share according to the number of males in each family (known as penis-counting), oth-ers according to the number of livestock in each. The economic gain or strain for any given individual therefore depends on the number of males or animals in his immediate family and the overall size of his dia-paying group. Since the latter may number in the thousands, an individual's share of the payment for an in-jury by a fellow member not in his immediate family may be al-most negligible.[32]

Dia-paying groups also handle conflicts between their own members according to terms specified in the membership agree-ment. In many groups, for instance, the compensation payment for killing a fellow member is lowered to the amount normally allotted to a family group directly involved in the killing of a nonmember (typically one-third of the total value), and only the victim's and offender's immediate families participate in the pay-ment. For the principals and their close kin, an injury within the dia-paying group thus has the same economic significance as an injury by or to a nonmember.

Not surprisingly, disagreements sometimes arise about how a particular case should be handled. When this involves different dia-paying groups, an *ad hoc* panel of arbitrators—without fam-ily ties to either side—convenes to decide the question. These panels have no coercive enforcement powers and so must rely on their own persuasive abilities, public opinion, and the willingness of the adversaries to make peace. When a disagreement involves members of the same dia-paying group, however, the arbitrators can enforce the terms of the membership contract coercively, even by permanent expulsion from the group itself. A man ex-pelled from his dia-paying group effectively becomes an outlaw, without protection and unable to prosecute grievances unless and until another group accepts him as a member.

Individuals not qualified by ancestry to join a dia-paying group are sometimes allowed to do so on a completely contractual basis.

Caravan traders passing through an area occupied by foreign clans often arrange protection of this kind, as do members of out-caste occupations (such as hairdressers and shoemakers) who do not belong to the major clans. Unprotected individuals tend to follow a Somali proverb warning everyone to "be a mountain or attach yourself to one." They attach themselves to a dia-paying group.

Dia-paying groups occasionally form temporary alliances with others representing the same lineage. In this way, 100,000 or more men may join together to wage a particular conflict, all bound by an agreement similar to those described above. The power of these alliances can be considerable, while the risk for each individual member—the share of a compensation payment that might have to be contributed—is extremely small in each instance. Even so, violence often escalates under these conditions, and the costs of compensation may correspondingly rise to such a level that the alliance loses its attractiveness. For routine conflicts, Somalis prefer dia-paying groups on a smaller scale.

In sum, virtually everyone in northern Somalia belongs to an organization, called a dia-paying group, designed to handle conflicts in everyday life. Lone individuals never face organized adversaries except when refusing to cooperate with their own dia-paying group. These groups not only pursue grievances of members against nonmembers but also assume liability for injuries inflicted by members on nonmembers. In principle, therefore, any injury is remediable with a payment of compensation, regardless of the wealth of the individual who is responsible. Every injury is incorporated by a group.

We next consider a social technology by which individuals in contemporary societies such as the United States might similarly enjoy the benefits of organization in everyday conflicts. This technology would radically change the social structure of the cases and provide some of the advantages of tribal life to people now impoverished in this respect. It would also profoundly alter the character of modern justice.

## Legal Co-operative Associations

Imagine a version of the dia-paying groups of traditional Somalia in a modern society such as the United States. Organizations of this kind, which might be called *legal co-operative associations* (legal co-ops), could be brought into existence with little effort and would almost certainly spread throughout society. They would transform the legal order as we know it, reducing organizational and other discrimination, punishment, and possibly even illegality itself. They would also encourage compensation, conciliation, and banishment as meaningful remedies in modern life.

### Jurisdiction

Legal co-ops would offer services partially resembling those of insurance and prepaid legal assistance programs, but their jurisdiction would be vastly greater. They would pursue grievances on behalf of their individual members and represent their members whenever and wherever grievances might be brought against them. They would collectivize the conflicts now defined and handled as the business of individuals. This process of incorporation could occur at any stage of a grievance, whenever an individual member might choose to notify a legal co-op that help is needed. These organizations would handle conflicts of any sort, regardless of their legal pedigree, and would be available around the clock.[33]

A tenant with a landlord problem, for example, could ask his or her legal co-op to handle the matter from the beginning. So could a landlord with a tenant problem. It would not matter whether the tenant and the landlord were members of the same or different legal co-ops, since each association would be available to handle conflicts internal as well as external to the membership. Like the Somali practice of dia-paying, moreover, compensatory damages or other payments would be received or provided primarily by the legal co-op, though the individual directly involved in each case would receive or contribute a disproportionate share. (Recall that in Somalia the immediate family involved in a case receives or contributes one-third of the

damages.) The economic incentive of individuals to bring grievances would thereby decline but not disappear, and so would
their liability.

Legal co-operative associations would intervene in both criminal and civil matters. If a member were arrested, for instance,
the legal co-op would provide a number of services, including legal and personal advice, an attorney, bail to obtain release before trial, and a fine that might ultimately have to be paid. It
would also assist the victim of a crime during the aftermath of
the incident. This might include contacting the police or prosecutor at the outset, helping the victim explain what happened,
arranging for medical care or other services (such as replacement of an automobile, lock, or window), and appearing in court
to describe the impact of the crime on the victim and to argue
for a particular disposition of the case.

The intervention of legal co-ops in criminal matters would increase the use of nonpenal modes of resolution now marginal to
the system of criminal justice. For example, the use of compensation could be greatly expanded, since a legal co-op would be
able to offer immediate financial satisfaction to a victim willing
to withdraw a criminal complaint against one of its members.
Surely numerous victims of burglaries, robberies, and other pred-
atory crimes would prefer a monetary settlement to prosecution,
particularly since conviction is never a certainty and victims
often find little consolation in the punishment of the offender
anyway. At present, hardly anyone is available to provide compensation in criminal cases.[34] The kinship system that traditionally accepted this liability has weakened or disintegrated, and
the vast majority of criminal offenders are too poor to provide
compensation themselves. Legal co-ops would thus advance the
evolutionary trend by which organizations are steadily assuming
the functions once performed by families.

Many crimes now handled by the police and courts might
never come to their attention at all, but would be handled from
beginning to end entirely by legal co-ops representing the principals. Besides increasing the use of compensation in criminal
cases, legal co-ops could facilitate reconciliations between adversaries, particularly in cases involving members of the same
family, friends, neighbors, or others with a prior relationship.

Whereas presently a crime and its prosecution tend to destroy any bond between the offender and victim, a legal co-op would be able to act as a mediator or negotiator to help the adversaries find a peaceful resolution of their conflict, possibly with a view to the preservation of their relationship. Conciliation would be especially feasible for offenders and victims belonging to the same legal co-op—which might have its own mediation or negotiation services as a privilege of membership—but conflicts involving different co-ops could be similarly handled as well. Legal co-ops would provide a new resource in the growing movement to increase the availability of mediation and negotiation in modern life.[35] By reducing the use of criminal justice while enhancing the role of compensation and conciliation, they would effectively restore patterns of conflict management found in many tribal and other simpler societies.[36]

Legal co-ops could also handle civil disputes in a conciliatory fashion. In fact, the distinction between criminal and civil cases would lose much of its contemporary significance. Legal co-ops would handle all cases in much the same way, with an emphasis on resolution rather than legal enforcement in a formal sense. This too is characteristic of simpler societies. In several respects, then, legal co-ops would retribalize the handling of conflict in modern life, providing members with benefits once available from families, clans, villages, and other primordial groups. Their benefits would be available to everyone, however, not merely to those fortunate enough to enjoy the attachments of kith and kin.[37]

*Membership*

Membership in legal co-ops could be voluntary or compulsory. If voluntary, it could be made universally available by means of subsidies, possibly governmental, to those otherwise excluded because of economic hardship. Although a large proportion of the population might choose to join a legal co-op (if only to avoid the disadvantages that nonmembers would noticeably suffer), some might choose not to belong, just as some nowadays choose not to carry property or health insurance. Compulsory membership could easily be arranged as well, such as by linking it to employment, like Social Security in modern America. Employers would then deduct the cost of "legal security" from each person's earnings. Those without employment could make other arrangements and, if indigent, could be subsidized. How many members legal co-ops would have, their social composition, how they could be financed, how their activities would be regulated, and other questions remain to be answered, but these details need not concern us here. It might be noted, however, that a high degree of social similarity across legal co-ops would seem essential to their primary function of equalizing conflict. Perhaps this could be adequately achieved by assuring that each would have a heterogeneous membership of about the same size as the others. Regional associations might allow this, for example, and each could be a local chapter of a national association that would monitor and regulate the entire system.

Another challenge would be to minimize social differentials in the handling of cases within and between legal co-ops. The risk of such differentials would seem greater than in the Somali dia-paying groups, where the memberships are more homogeneous. Equal treatment might require, for example, that the social characteristics of each member be as invisible as possible to the administrators of each co-op (see Chapter 4 for a discussion of this general strategy for reducing discrimination).

Still another matter for consideration is how legal co-ops would handle their recidivists, i.e., those who might repeatedly injure or offend others and thereby incur disproportionate costs (financial and otherwise) for their fellow members. Such individuals could readily be subjected to social pressure of various kinds: Their names and misdeeds could be publicized among the membership, for instance, and their reputations threatened or tarnished. Troublemakers could also be subjected to added surveillance. Legal co-ops would thus constitute a new medium of social control in modern life and, at the same time, would revive practices that largely disappeared with the tribes and villages of traditional society.

Reputations presently exist in only a few limited situations, such as the workplace and family. Most misconduct is unknown beyond those immediately involved, and the surveillance of dangerous people is left almost entirely to the police. Legal co-ops would significantly reduce the anonymity of recidivists.

They might also apply another Somali practice: Wrongdoers could be required to contribute disproportionately to the costs of their misconduct. This would place a growing burden of debt on repeat offenders and possibly their immediate families. But despite deterrents such as these, some individuals would undoubtedly persist in recurrent misconduct beyond the tolerance of their fellow members.[38] What then could be done? This is a fundamental problem in any system of collective liability: How can individuals be held responsible for their conduct, and how can they be encouraged to feel responsible for it, if a collectivity suffers most of the consequences for their actions?[39] Why should anyone bother to behave?

Tribal societies long ago found a simple solution to recidivism when all else fails: banishment. Social control internal to the group invariably accompanies collective liability, and the ultimate sanction is expulsion:

> Where every member of a corporate group has the power to commit it to a collective liability, a corollary rule always exists whereby the corporation may discipline, expel, or yield up to enemies members who abuse this power or whom the corporation does not choose to support in the situation in which he has placed them. . . . Expulsion is a qualifier of collective liability.[40]

Recall that Somali dia-paying groups may permanently expel members who are uncooperative. Legal co-ops could similarly expel repeat offenders, reinstating a punishment often thought obsolete or even impossible in modern life. Banishment would be a particularly meaningful sanction if other legal co-ops were to refuse membership to the same individuals.[41] A society-wide blacklist of recidivists would create a new class of outlaws, stripped of the advantages of organization and vulnerable to all.[42] They would be doomed to individuality.

In recent years, legal scholars and social critics have often drawn attention to the advantages enjoyed by organizations when they engage in illegality, particularly when they victimize individuals.[43] It has even been suggested that "the central problem of modern civilization is the control of organizational life."[44] Many reforms have been proposed, but all are variations on a single theme: They seek to reduce the advantages of organizations by increasing their vulnerability to law, whether by increasing governmental surveillance of their activities (such as by requiring more internal record-keeping and policing) or by handling their misconduct with greater severity (such as by stiffening penalties and broadening liability).[45] It is also argued that we should handle deviant organizations like ordinary criminals, such as by placing them on "probation" when appropriate[46] or even, in extreme cases, by "capital punishment"—by revoking their legal right to exist.[47]

All these proposals, however, have one shortcoming: They are impossible. Just as physical impossibilities cannot be wished into existence, such as objects that violate the law of gravity by falling up instead of down, so there are sociological impossibilities that violate the principles of social behavior. An example would be a legal system without social differentials in the handling of socially diverse cases—provided, of course, that it resembles the legal systems of the past in other respects. A legal system in which organizations have no advantages over individuals is similarly impossible. These advantages are consistent with existing sociological theory,[48] and no legal system without them has ever been observed. To propose their elimination, simply by mandate, is much the same as proposing that henceforth all objects shall fall up instead of down. It is unscientific. Wishful thinking. Fantasy.

The idea of legal co-operative associations is different. Unlike earlier proposals that we should suddenly rescind the legal advantages of organizations and begin treating them like individuals, legal co-ops provide a social technology that gives individuals the same advantages as organizations. Legal co-ops would effectively transform individuals into organizations for purposes of legal action, altering the social structure of cases, equalizing the amount of organization on each side. Countervailing the power of organization with organization itself, they would radically reduce organizational discrimination in any legal system. Moreover, the idea of legal co-ops applies rather than contradicts sociology. It conforms to everything we know about the natural behavior of law.[49]

To conclude: Legal co-operative associations would incorporate many conflicts now wholly the concern of individuals and replace individual victims and offenders in a vast range of situations. Conflicts are property,[50] and legal co-ops would collectivize their ownership. Available 24 hours a day, they would intervene, financially or otherwise, at any stage of any conflict. They would significantly change the management of conflict in modern societies such as the United States, undermining the advantages of organizations over individuals, substituting compensation and conciliation in many cases presently punished as crimes, even restoring banishment as a solution to the problem of recidivism. Legal co-ops would balance and homogenize the social structure of conflict, obliterating many of the social differences between cases and reducing discrimination of all kinds. They would raise the social stature of nearly everyone.

Modelled on the dia-paying groups of Somali nomads, legal co-ops would retribalize law and conflict in today's world.[51] But they would be available to all, and not merely to those with families or other primordial allies. They would bring law into the mainstream of social evolution, providing organization to individuals with legal problems, among the last unorganized people in modern society. Legal co-ops illustrate how legal sociology can be applied to a practical problem: Change the social structure of the cases, and you will change how they are handled. Equalize them, and you will equalize justice.

# 4

# *The Desocialization of Law*

A leading scholar in legal sociology once proposed that the "chief preoccupation" of the field should be "the rule of law."[1] The main challenge, he suggested, is "to discover which social conditions are congenial to the rule of law and which undermine it."[2] And, indeed, insofar as "the rule of law" refers to the universalistic application of legal rules to human affairs,[3] it is now possible—20 years later—to specify when these conditions exist: never.

A great deal of evidence, reviewed in Chapter 1, reveals dramatic variation in the handling of technically similar cases in a wide variety of societies and legal settings. This evidence casts doubt on the possibility that a legal system such as those that have existed in the past will ever achieve universalism as a routine mode of operation. Rather, as long as the cases sociologically resemble those that have always been seen, and as long as they are always processed by the same procedures, it appears that there will always be social differentials—discrimination—in how they are handled. In the absence of legal reform, universalism in the application of law would seem a utopian dream.

The previous chapter introduced a technique designed to reduce discrimination by socially restructuring the cases. We could create organizations specifically to handle conflicts—legal cooperative associations—establish them throughout society, and make them available to everyone. Legal co-ops would not only lessen the vulnerability of individuals to organizations but would also socially homogenize the cases to a significant degree. Organizational and other discrimination could thus be counteracted and neutralized. This chapter describes another technique to reduce discrimination. It shifts attention from the social structure of the cases to the process by which they are handled, illustrating a second strategy by which a legal system might be sociologically reformed. Before this technique is discussed, however, we further explore the nature of legal discrimination and the conditions under which it occurs.

## The Quantity of Discrimination

Although variation in the handling of technically similar cases is ubiquitous in legal life, its extent is not everywhere and always the same. The amount of this variation itself varies across legal settings of all kinds. It differs across societies, communities, courts, and cases.

In modern America, for example, the amount of variation in the handling of homicide cases is spectacular, ranging from those that legal officials decide not even to investigate (as frequently occurs when prisoners or skid-row vagrants kill each other)[4] to those resulting in capital punishment (as may happen when a poor robber kills a prosperous stranger). Numerous possibilities lie between these extremes, such as when an investigation results in a strong case but a grand jury refuses to indict the defendant, when a judge or jury refuses to convict the defendant despite overwhelmingly incriminating evidence, when a conviction results and the defendant receives probation alone, only months in confinement, several years, a number of years, many years, and so on.[5] This immense range of variation contrasts with the seemingly mechanical reaction to other offenses—where the penalty for each is largely the same for all—as, for example, in the handling of public drunkenness, prostitution, and parking violations.[6]

Variation in the handling of homicide cases, however, is

greater in some societies than others. In the tribal societies studied by anthropologists, for instance, the amount of variation is considerably less than in more complex societies.[7] And some modern societies, such as the European nations, have less variation than others, such as the United States and South Africa.

What explains these differences? Why does the amount of legal variation itself vary to such a degree? When does discrimination increase or decrease? The answer is simple.

### Social Diversity

Insofar as the social structure of a case predicts how it will be handled (as outlined in Chapter 1), the extent of structural differences among and between cases will determine the extent of legal variation across them. If different degrees of intimacy divide the adversaries from one case to the next, for example, or different levels of social status, or if the legal officials handling them differ in their social characteristics, the legal outcomes will vary to a greater extent than if the cases are more simliar. In other words: *Legal variation is a direct function of social diversity.*[8]

This is why the handling of American homicides varies so much: There is enormous diversity in who kills whom, ranging from cases involving the poorest and most marginal individuals to those involving the extraordinarily wealthy and powerful, and from those involving husbands and wives or parents and children to others involving friends, neighbors, workmates, cellmates, acquaintances, or complete strangers. They might involve whites, blacks, Asian-Americans, Chicanos, Puerto Ricans, Jews, or many other possibilities in any conceivable combination. Homicides in America are as socially diverse as America itself, even if most occur at the bottom of society between family members, lovers, or friends.[9] The degree of social diversity across homicides is not nearly so great in Europe, and so the range of variation in the handling of these cases is considerably narrower. Capital punishment declined and disappeared from European societies with the decline of upward killings across great distances in social space—those committed by people of lower status against socially superior and culturally alien strangers. As long as the poor continued to kill the rich of Europe, such as during robberies or rebellions, capital punishment continued as well. But these killings are now rare. In America, however, upward homicide across great distances in social space still occurs quite frequently, such as when poor blacks rob and kill white businessmen, and (as noted in Chapter 1) these are the cases especially likely to receive capital punishment. Tribal societies are even more homogeneous than European nations, homicidally and otherwise, and so their handling of killings displays still less variation. Where everyone herds livestock, for example, differences in wealth—animals—are often small and unstable. When this is so, all killings are handled in much the same fashion: The victim's family typically demands and receives a standardized payment of livestock.[10] Differences in the degree of intimacy between the killer and victim are the major source of whatever variation occurs. Among the Nuer of Sudan, for instance, the number of cattle expected as compensation decreases (or is totally absent) when the killer and victim belong to the same family.[11]

The degree of variation in the handling of each category of crime depends on the social diversity among the alleged offenders and victims. In the United States, for example, public drunkenness and prostitution cases tend to be socially similar from one case to the next, and so they are handled similarly as well. In fact, the handling of all so-called crimes without victims (such as gambling and liquor-law violations) exhibits less variation than crimes with identifiable victims. This is partly because of the sociological relevance of the victims themselves: Crimes with victims are more socially heterogeneous both because of the social diversity among those selected as victims—combined in various ways with the offenders—and because the social relationship between the victim and offender introduces its own dimension along which cases diversify. Are the victim and offender acquaintances, strangers, or what? Crimes without victims are not similarly variable, and so neither are their legal destinies. Finally, variation in the handling of crime also arises from social diversity across third parties such as judges and juries.

*The Relevance of Race*

One variety of legal variation in modern America is known as racial discrimination. Although strictly speaking a biological rather than a sociological trait, race in the United States is systematically associated with a number of social conditions. More blacks than whites are poor, unemployed, and uneducated, and fewer blacks than whites own their own home, live in a two-parent family, or enjoy social resources of any kind. Blacks therefore suffer legal disadvantages—the fate of anyone, of any race, inhabiting the bottom of society.[12]

Homicide cases again provide an instructive example. As noted in Chapter 1, an American homicide is most likely to result in capital punishment when a black kills a white, next when a white kills a white, then when a black kills a black, and it is least likely when a white kills a black.[13] Since most homicides (and other crimes) occur between members of the same race,[14] most blacks convicted of homicide are killers of fellow blacks and receive comparative leniency. But since more blacks victimize whites than vice versa, a pattern that attracts the greatest severity, the overall severity toward each race tends to average out. If we examine only the race of the defendant, then—ignoring the victim's race—the evidence misleadingly suggests that racial discrimination does not exist.[15] Discrimination against blacks becomes visible only when we hold constant the social characteristics of the victim: Blacks who victimize blacks receive harsher treatment than whites who victimize blacks, and blacks who victimize whites receive harsher treatment than whites who victimize whites.

In a recent study, however, the victim's characteristics were held constant and yet no evidence of discrimination against blacks was found.[16] This study provides uniquely important evidence about the legal relevance of race. The investigator expressly sought to examine racial discrimination in court decisions, but after an exhaustive and sophisticated statistical analysis he concluded that race had no effect whatsoever on the handling of the cases. Because his results seemed inconsistent with previous research, he was completely baffled. How could this happen? The location of the study explains the results. Unlike past research on discrimination in American law, this was a study of *military* justice. It examined how the U.S. Army handles individuals charged with being absent from an assigned post without permission (known as absent without leave, or AWOL). Since the "victim" is always the same—the United States itself—the investigation could focus entirely on the effect of the defendants' characteristics: Are blacks treated more severely? The central finding was that the decisions of military courts are predictable from the offenders' conduct alone, such as how long they were absent without leave or whether they have committed other offenses in the past. Race does not matter. It has no statistical significance. But this finding should not actually be regarded as anomalous.

The legal irrelevance of race is understandable in light of the study's setting—the U.S. Army. In the Army, everyone of each rank has essentially the same social existence as everyone else, regardless of skin color. They have the same employment status, occupation, income, residence, and the same round of activities. They even dress the same. In other words, the military personnel of each rank—and military defendants—are extremely homogeneous. It is unlikely that any civilian court of law in modern America handles blacks and whites so equal in their social characteristics. Hence, it is unlikely that blacks and whites are treated so equally anywhere else.[17] Military justice shows that legal discrimination disappears in the absence of social diversity.

## The Jurisprudence of Parking Tickets

Trivial as it might first appear, the enforcement of parking regulations also has unique relevance to the understanding of legal discrimination. This is because, like military justice, it has a seemingly anomalous feature that cries out for explanation: little or no discrimination. If law varies across social space, as legal sociologists repeatedly demonstrate, why do police officers typically handle illegally parked automobiles in such a mechanical fashion, case after case, without significant variation? The usual practice of officers checking parking meters, for example, is to walk from one to the next, writing a citation (ticket) for every violation, without exception. The drivers of these automobiles differ greatly in their social characteristics, and yet discriminatory treatment does not occur. Why not? Why does the rule of law prevail among parked cars?

The answer is that parking violators have an attribute rarely if ever found among anyone else subjected to the legal process: They are socially invisible. No matter what the social characteristics of automobile drivers might be, no matter how much they differ among themselves, when they violate parking regulations their social identities are largely unknown. They are processed *in absentia*. The police see only the automobiles, and in most cases these reveal little about the social characteristics of their drivers.[18] In this sense, parking justice is truly blind.

Of course, parking violators are socially invisible only when the police are unaware of who operates or owns illegally parked automobiles—a state of affairs found primarily in urbanized and other impersonal settings. In small towns and villages, the police often recognize the violators from their vehicles, and so more variation in enforcement undoubtedly occurs.[19] In urban settings, the handling of violations by moving vehicles, such as speeding or disobeying traffic signals, involves more variation as well, since in these cases the police meet the drivers and learn a few of their social characteristics. Some years ago in one American city, for example, officers openly behaved with greater severity toward black and teenaged drivers, and they granted special immunity to parents accompanied by children.[20] Similarly, an experiment in California revealed that the police were more likely to stop and ticket vehicles bearing a bumper sticker of the Black Panther Party, a radical political group.[21] On the other hand, police officers themselves are effectively immune to the traffic laws throughout the United States. If stopped for a violation in their private automobiles, officers need merely identify themselves as such to be exonerated. But when they illegally park their private vehicles, police officers are indistinguishable from other violators and are unwittingly treated like everyone else.

## Law and Social Information

The absence of discrimination in parking enforcement shows the legal fatefulness of information about social characteristics, or *social information*.[22] Where no such information enters a legal process, technically identical cases are socially indistinguishable and, regardless of the actual diversity among them, will be treated the same.[23] Social characteristics are relevant only if they are known. The universalistic enforcement of parking regulations thus illustrates another principle of legal sociology: *Legal variation is a direct function of social information*.[24]

Legal discrimination of every kind—racial, economic, cultural, organizational, etc.—depends on the amount of social information entering each legal setting. Even if the cases are socially identical, different amounts of social information about each will make them appear socially different and introduce variation in how they are handled. If, for example, the defendant's criminal record (normative information) comes to the attention of a judge or jury in only some instances, only these defendants will be disadvantaged.[25] The same applies to economic information (about their wealth), cultural information (about matters such as their ethnicity and religion), relational information (about their intimacy with the victim), and so on. Every aspect of the social structure of cases may become more or less visible and thereby have more or less impact on how they are handled. Social information is an essential ingredient of discrimination.

### Social Information as a Quantitative Variable

The quantity of social information varies considerably from one legal setting to another. In the traditional societies of Africa, Oceania, and South America, for example, the participants know all the social characteristics of the cases arising in a village or camp virtually all the time. Everyone is well acquainted and need not even inquire about each other's social characteristics. Far less of this information is available in the legal settings of modern societies, but some have more than others. Legal officials in the Soviet Union, for instance, seem to gather social information about defendants more methodically and extensively than do Western officials.[26]

Different amounts of social information also appear at each stage of the legal process. The earliest stage of the criminal process—a telephone call to the police—contains the least. The individual answering the call learns the nature of the problem and its location (possibly revealing the status of the neighborhood involved), the caller's name may be given (possibly revealing an ethnic identity), and the caller's speech may indicate other characteristics. The caller is also likely to mention whether the victim is a business or other organization. Nothing more, however, is normally learned. Because the police have so little social information about these calls, there is little variation in how they are handled: A police car is dispatched in nearly every case in which help is requested.[27] Even so, the alacrity with which this occurs may vary with the few shreds of social information communicated, so that, for instance, the police may give faster service to a suburban household or business firm than to someone calling from a slum. The officers dispatched might also respond more or less promptly according to the same factors.[28]

Social information expands greatly when the police arrive at the scene of an incident, since there they meet the people involved and may inquire about some of their social characteristics not otherwise obvious (such as their occupations and the degree of intimacy between them). After police contact, therefore, variation in the handling of cases increases significantly.[29] If the police make an arrest and the case reaches criminal court, more information about all concerned enters the process and produces still more variation. Finally, if a conviction occurs, more information about the defendant's social history may be uncovered before the final disposition. An official may prepare a so-called presentence report, for example, detailing the defendant's criminal, school, employment, military, family, and residential history. This results in still more variation, some cases receiving suspended sentences or probation, others longer or shorter sentences to jail or prison.

Some cases generate more social information than others, partly as a function of the maximum penalty at stake. Capital cases and other felonies generate more social information than misdemeanors, for example, which in turn generate more than traffic offenses.[30] Some lawyers and prosecutors also introduce more social information about everyone involved—victims, witnesses, defendants—than others. It might be added that anyone practicing "sociological litigation" (as outlined in Chapter 2) would intentionally seek to alter the balance of advantage in this fashion.

## The Decline of Social Information

Social information varies not only across legal settings but across social settings of all kinds. For example, a commercial transaction may occur between people who know one another's social characteristics, as in business relationships in small towns and villages, or it may lack this dimension almost entirely, as in business by mail or telephone between complete strangers. The same applies to professional transactions. Medical care might be provided by a family doctor who knows the patient's entire social history, for instance, or in the emergency room of an urban hospital where patients are nearly indistinguishable. Similarly, in some neighborhoods and apartment buildings the residents learn about each other's occupation, marital status, ethnic background, religion, etc., while in others they learn almost nothing. And just as legal variation depends on this information, so does social variation of other kinds. Some people may get better prices, better credit, better health care, better life conditions of one sort or another. Discrimination can occur anywhere. But these differences can arise from social factors only if they are known. All discrimination requires social information. We can also see an evolutionary trend in the quantity of this information throughout society: It is decreasing.

Over the centuries, information about social characteristics has been steadily attenuating in all walks of life. In this sense, human life is *desocializing*. Once everyone had abundant information about everyone else in their daily lives, but no longer. The telephone operator knew personally those who made calls; the doctor knew the patients; the banker, the grocer, and the tailor knew the customers. But now this is rare. The increased scale, organization, and fluidity of modern life have resulted in a desocialization of human affairs. So have electronic communications. Transactions of all kinds are increasingly standardized and impersonal. More and more, everyone is treated the same everywhere. The disappearance of social information is reducing discrimination throughout society.

Law, however, lags behind. It remains saturated with information about the social characteristics of litigants and others involved in legal affairs. Some cases, such as those resulting in a major criminal trial, are veritable feasts of social information about all concerned—with revelations of financial and family history, personal habits, associations, and improprieties. The abundance of this information, particularly in court, makes legal discrimination possible. But it could be reduced. Easily.

## The Desocialization of Courts

The drift toward a desocialized society has already routinized and depersonalized legal life to a significant degree. In large cities, for example, police officers, prosecutors, judges, and juries typically are unacquainted with the citizens they encounter in the course of their duties. They learn the social characteristics of these individuals only by visual and verbal cues and by direct inquiries. This contrasts with small towns and rural areas, where "everyone knows everyone else's business" and much information about those involved in legal matters is readily available.[31] Even in the largest cities, however, social information accumulates as cases progress through the several stages of the legal process. The technical description of each, contained in the original charge or complaint, gradually acquires social details concerning precisely who has a grievance against whom, who sides with whom, etc. In this sense, each case is socialized. And since the socialization of cases intensifies as they proceed through the stages of the legal process, so does the degree of variation in how they are handled. The most occurs in courts, particularly in final dispositions. The more social information, the more discrimination. It follows that a reduction of this information about every case—a desocialization of law—would reduce discrimination in legal life. A total desocialization of the legal process would totally eliminate legal discrimination.[32]

If only because lawyers and legal officials inevitably learn some of the social characteristics of cases when they first come to their attention, a total desocialization of law would be impossible.[33] A desocialization of the courts, however, is entirely feasible.

### Partial Desocialization

Social information in court could be reduced by new procedural rules prohibiting testimony and other presentations about the social characteristics of cases. These rules would list the excluded categories of social information, such as the race, ethnicity, and wealth of the parties involved in a case, their educational, occupational, marital, and residential history, and the nature of their relationship with each other. (To exclude ethnic information, the avoidance of proper names would also seem necessary.) All this information could be barred from courtrooms without otherwise altering their proceedings. In fact, American courts already desocialize their hearings to some extent. Information about social characteristics may be excluded from testimony as immaterial to a case, and so-called shield laws exclude particular details, such as testimony about a rape victim's sexual history. The same technique, addressed to a broader range of social information, would substantially reduce variation in the handling of cases.

Another technique would be to change the judge or jury for the final disposition of cases, so that the sentencing of convicted criminals or the awarding of civil damages would be assigned to individuals entirely unaware of any social information surfacing during the trial. This would eliminate social differentials in sentencing and the awarding of damages despite the contamination of trials by social information. Racial differentials in capital punishment, for example, could readily be ended in this fashion. But we can also imagine more extreme degrees of desocialization that would drastically curtail or even totally eliminate discrimination from the courts.

## Radical Desocialization

No matter how much new procedural rules might exclude social information from courtroom presentations, another factor would inevitably reveal a significant number of social characteristics of the parties: their mere presence. Physical appearance alone communicates a great deal about people, including their race, gender, and age. In addition, clothing, jewelry, hairstyle, and grooming suggest other facts about their social location, such as their probable social class and lifestyle. Another source of social information is speech. A person's choice of words, grammar, and pronunciation indicates much about social background and position. So does the style of presentation, such as the degree of assertiveness and verbosity. Its social meaning may not be consciously recognized, but speech behavior can limit or enhance the credibility of testimony.[34] Personal presentations alone thus bring social information into the courtroom. For this reason, a thorough desocialization of the courts would require more than the exclusion of particular subjects from testimony and other presentations. It would require the exclusion of people.

To eliminate the information about social characteristics conveyed by physical appearance and body decoration, it would be necessary for adversaries and witnesses to remain beyond the view of judges and juries.[35] Their testimony would have to be elicited earlier, as in present-day pre-trial depositions, and submitted to the court by their attorneys. To eliminate the social information conveyed by speech, all testimony would have to be communicated in written transcripts. (Electronic reproductions would allow socially revealing speech behavior to be heard.) Even written testimony would need to be edited to standardize language and style of expression as much as possible.

The removal of the adversaries and witnesses from the courtroom would mean that judges and juries could no longer observe their so-called "demeanor"—facial expressions, signs of nervousness, confidence, sincerity, etc. Legal experts regard demeanor as a major indicator of the veracity of courtroom testimony, and its elimination would undoubtedly hinder fact-finding to some degree. But the elimination of demeanor would reduce the role of social factors in court as well. As implied in Chapter 1, witnesses of higher status often have a demeanor that lends greater credibility to their testimony: Their speech is more fluent, precise, and forceful—more "powerful"—and for this reason alone they are more convincing.[36] Mere status in itself, if visible, also enhances credibility.[37] Since those of social standing benefit most from displaying themselves and their demeanor, any loss of truth entailed by removing witnesses from courtrooms might be justified as a cost of reducing social inequalities.

The removal of witnesses from the courtroom also would eliminate direct confrontations between accused persons and their adversaries, long considered a fundamental right of defendants. But this may be a dubious advantage for many defendants, particularly in criminal cases where they are nearly always socially inferior to their accusers. However valuable it might be in other respects, a personal confrontation in court draws attention to the social handicaps of some individuals and weakens their chances of winning. Since criminal defendants nearly always lose anyway,[38] confrontations with their accusers do not seem to help them a great deal. Furthermore, even if witnesses were barred from the courtroom, defense attorneys (or the defendants themselves) could still cross-examine hostile witnesses in closed-door sessions. Desocialization as described above would end only personal confrontations in open court where they facilitate discrimination.

A desocialized trial might theoretically include oral arguments by lawyers on each side. On the other hand, since the social characteristics of lawyers have their own legal consequences,[39] a still more complete desocialization of the courts would exclude these participants as well. Their arguments could be added to the other transcripts submitted to the court for its deliberations. Trial skills would become obsolete.

## Electronic Justice

Yet even radical desocialization would not totally eliminate discrimination from courtrooms. Another source would remain: judges and juries. Their decisions reflect their own social characteristics, introducing differentials of another kind.[40] The result is a disadvantage for defendants unfortunate enough to have their cases decided by judges and juries inclined toward severity and also for prosecutors, victims, and plaintiffs whose cases are decided by those inclined toward leniency. Although juries might conceivably be abolished, judges present more of a challenge in sociological engineering. Judgments are indispensable—the essence of law itself—and yet judges differ among themselves in their social characteristics and, hence, in how they handle cases. What might be done to eliminate this final source of discrimination? Can we have justice without judges?

However far-fetched it might seem today, the time may come when computers can be programmed to process complaints and testimony and to select dispositions. Universalistic treatment could then be achieved simply by using the same program for all cases.[41] This would accomplish the final step in the desocialization of courts: closing the courtrooms themselves. Law would reach a new evolutionary stage, and trial by judge and jury would join trial by ordeal and trial by combat in the ancient history of law. Unless we have already attained the highest level of legal evolution that will ever be reached, trial by computer may someday come to pass. Access to justice would become merely a matter of access to a computer. The capacity of the legal system would expand almost limitlessly, ending the old problem of delay in court. The cost of litigation would plummet.[42] Discrimination would disappear.

Electronic justice would be socially blind, but every aspect of legal life could not be handled in this fashion. Total computerization would be impossible. The discretion of citizens to pursue some grievances and not others would still introduce social differentials in the application of law. Police, too, would undoubtedly continue to select only some technically criminal cases for legal attention. Even so, other police work could be significantly desocialized. Calls from citizens might be electronically screened and patrol cars automatically dispatched, for instance, and at least some arrests and traffic citations might be decided by reference to a computer programmed to consider only the technical features of cases, including extenuating circumstances (such as speeding in a medical emergency). Prosecution decisions might be similarly desocialized and standardized. So might numerous other legal decisions, such as those of regulatory agencies and parole boards. No matter how much the handling of cases might be desocialized, however, a major source of discrimination in the application of law would survive: the definition of illegality itself.

Even if common crimes such as burglary and robbery were handled by a totally desocialized process, without any social information whatsoever about the defendants or their alleged victims, the same segments of the population would continue to be imprisoned: the poor, the unemployed, the young, members of minority groups, and other disadvantaged people. Since these people commit nearly all the common crimes, they would continue to be the primary ones convicted on these charges. The desocialization of courts would eliminate variation in the probability of conviction and in the severity of punishment arising from social differences, but it would not and could not change the selection of the defendants themselves. Apart from an occasional case involving a middle-class ("white-collar") defendant, the jurisdiction of criminal law would continue to apply overwhelmingly to people at the bottom of society.

It does not matter that anyone can, in principle, choose to commit a crime. The fact is that only some are likely to do so: those at the bottom. Realistically speaking, only their conduct is defined as criminal. As a sarcastic observer once remarked, "The law in its majestic equality . . . forbids the rich as well as the poor from sleeping under bridges, begging in the streets, and stealing bread."[43] Legal rules might in theory apply to everyone, and courts might someday handle everyone equally, but legal authority will always and everywhere fall unevenly across social space. Discrimination of this kind is inherent in law itself.

# 5

# *The Delegalization of Society*

People in modern society often try to eliminate social differentials in legal life by wishing them away. "Discrimination is prohibited," they declare. "All cases must be treated the same." Such pronouncements typically seek to equalize the treatment of socially different cases by imposing equally *harsh* treatment on all. In so doing, they obey a long tradition of legal reform seemingly based on the following principle: To improve law, add more law. To combat organizational discrimination, for example—the legal advantage enjoyed by businesses, governments, and other corporate beings—reformers virtually always propose that organizations and their representatives be subjected to more rules and regulations, more surveillance, more prosecutions, more lawsuits, and more severe penalties.[1] Similarly, evidence of leniency toward "white-collar" criminals inspires demands for more severity,[2] as does evidence that American police handle assaults in families more leniently than those involving strangers.[3] Legal reformers prefer equal severity to equal leniency.

The reforms outlined in Chapters 3 and 4, however, do not seek to reduce legal discrimination by changing the amount of law. Nowhere does this book suggest that legal officials could or should suddenly begin handling socially different cases in the same fashion, whether severely or leniently. Instead, these reforms would reduce or eliminate the social differences between the cases themselves. One would homogenize the cases by giving ordinary citizens the advantages of organization (legal co-ops), while the other would homogenize them by making their social characteristics invisible (desocialization). Neither would attempt to create a legal system where cases known to be socially different are treated the same. This would appear to be sociologically impossible. Instead, both reforms assume that law will continue its present role in the conflict management of the future, that it will be as available as always and will have the same penalties and remedies as always. Both reforms also assume that the definition of illegality will remain the same, even though (as noted in the last chapter) this too is a source of discrimination.

Now we relax our assumptions and consider a final technique by which legal discrimination might be reduced: the reduction of law itself.

At least since Thomas Hobbes argued centuries ago that a society without a powerful state would inevitably have a "war of every one against every one,"[4] it has been widely believed that law is essential to a civilized way of life. Even so, a great deal of anthropological evidence shows that stateless—and lawless[5]— societies have existed in profusion without the reign of terror envisaged by Hobbes and others.[6] People have actually lived without law for nearly all of human history: It was invented only about ten thousand years ago, when agriculture began to replace hunting and gathering as the primary mode of subsistence.[7] Furthermore, even in legalistic societies such as modern America, the vast majority of conflicts are handled without law.[8] What, then, do people do without it? How do they achieve justice?

## Alternatives to Law

Today as in centuries past, people with grievances may select any of several modes of conflict management: (1) self-help, (2) avoidance, (3) negotiation, (4) settlement by a third party, or (5) toleration.[9] Every society has its own configuration of these strategies, but all are found everywhere to some degree.

## Self-help

Self-help is the expression of a grievance by unilateral aggression. It might entail direct criticism, ridicule, harassment, the destruction of property, banishment, or violence, including homicide. One of its several scenarios is vengeance, where someone tries to "even the score" against a wrongdoer, illustrated in the biblical notion of "an eye for an eye."[10] Feuds involve reciprocal vengeance over an extended period of time.[11] Another scenario is the defense of honor, which often has a ritualistic or game-like character and arises in response to insults as well as physical injuries. Honor is akin to "manliness," and a display of courage, even if disastrous, still protects the honor of an aggrieved individual.[12] A third scenario of self-help is discipline. Here the grievance is handled hierarchically, such as when a slave is punished by a master, a prisoner by a guard, a student by a teacher, or a child by a parent.[13] Still another scenario is rebellion, where the grievance arises from below, such as when a slave, serf, or other underling violently attacks a despised overlord.[14]

Many simple societies with herding and horticulture have a great deal of self-help.[15] It is also favored by aristocrats and peasants in feudal societies and by lower-class men, teenaged boys, and prisoners in modern societies.[16] Much occurs in international relations as well.

## Avoidance

A second mode of conflict management is avoidance, or the curtailment of interaction. It may be initiated by an aggrieved party, an offending party, or both at the same time. It may be total or partial, permanent or temporary, and involve physical separation or only a reduction of contact or communication. Examples are the "cold shoulder," resignation from an organization, divorce, desertion, migration, and suicide. People in hunting and gathering societies frequently employ avoidance,[17] as do disgruntled consumers,[18] business executives,[19] and suburbanites.[20] Avoidance has been a major alternative to law throughout human history. Some archaeologists believe that law itself emerged historically only when populous societies evolved in limited spaces where avoidance was difficult.[21]

## Negotiation

Another alternative to law is negotiation, a process in which people discuss their conflict and seek a resolution, usually a compromise. Negotiation occurs in a wide variety of social settings[22] and is a primary mode of conflict management in many traditional societies.[23] American lawyers employ negotiation to assist those accused of crime and to settle civil disputes in the vast majority of cases.[24] Hence, the enormous increase of lawyers in recent American history—a doubling during the past 25 years[25]—suggests a growing demand for negotiators.

## Settlement

A fourth mode of conflict management is settlement by a mediator, arbitrator, judge, or other third party.[26] Law, usually a form of adjudication, occurs under comparatively rare conditions (see Addiction to Law below). Mediation, where a third party acts as a broker between the adversaries, is more prevalent historically, perhaps because it requires no coercion and readily oprates in nongovernmental settings.[27] Undoubtedly the most common third-party behavior, however, is gossip. A kind of trial in absentia, less intrusive and authoritative than other forms of settlement, gossip plays an important role in many groups.[28]

## Toleration

The fifth mode of conflict management may appear to be nothing at all, but when a grievance is present, doing nothing is a response in itself. Known colloquially as lumping it, turning the other cheek, learning to live with it, and biting the bullet, toleration is probably the most frequent response to conduct regarded as wrong, improper, injurious, or otherwise deviant.[29] But toleration is more likely in some social locations and directions than others, such as when the offender is socially superior to the aggrieved party or when both are socially marginal or unrespectable.[30]

Most illegality is tolerated. That is, most does not result in legal action. In modern America, for instance, citizens notify the police of only about one-third of the incidents they regard as "crime."[31] If we include technically criminal cases not so regarded by citizens (such as most violence in families and between friends), the proportion of crimes reported is still smaller.[32] People appear even more likely to tolerate civil problems. According to one study, aggrieved parties contact a lawyer in only about one-tenth of the cases involving a potential claim of $1,000 or more.[33] When they do contact a lawyer or legal official, most often the case ends there, without further action, though technically it could go to court.[34] In addition, a court decision against a defendant does not necessarily mean that the full force of law will be applied. Numerous convicted criminals are released on probation or suspended sentences, for example. Permissiveness may be an effective way to avoid a "vicious circle" whereby the punishment of offensive behavior leads to frustration and hence more offensive behavior.[35] But however justified, everyone practices toleration to some degree.

## Legal Overdependency

If law is a relatively unusual form of conflict management, how should we understand its great proliferation in modern times? Why has it come to overshadow the many alternatives preferred for most of human history? What causes it?

### Addiction to Law

Sociologists have identified a number of social conditions that encourage the growth of law, including inequality of wealth, social atomization, the division of labor, the presence of organizations, and cultural heterogeneity.[36] Another such condition has a practical relevance we explore in this chapter: the absence of alternatives.

People have not necessarily come to prefer law to other modes of conflict management; rather, increasingly nothing else has been available. When people have a grievance and law is the only available means of redress, they use it more readily than otherwise. Stated more precisely: *Law varies inversely with other social control*.[37] One of the most powerful principles of legal sociology, this pattern appears in the evolution of law over the centuries as well as the case-by-case handling of grievances. Historically, law expanded its jurisdiction as the family, village, and other communal formations declined,[38] and, in everyday life, people resort to law more willingly when other modes of conflict management are scarce or absent. The fewer alternatives, the more law. And the more people resort to law, the more they come to rely on it. They develop a condition of legal dependency.[39] In this sense, law is like an addictive drug.[40]

Totalitarian societies, such as Stalin's Soviet Union and Hitler's Germany, seem to have reached the highest level of legal dependency ever achieved. Moreover, a recent study indicates that these regimes achieve their nearly total control primarily by encouraging people to bring all their grievances, no matter how petty, to state officials. This is most easily and effectively done by denouncing enemies to the political police. The result is immediate (imprisonment, exile, or worse), even when evidence of wrongdoing is absent and the actual reason for a case is personal (such as a grudge against a workmate, landlord, or neighbor). Ordinary citizens unwittingly draw the state's authority into all walks of life:

> The rank-and-file members of society have a relationship to the state similar to that of the state's functionaries; that is, they can and do use the state freely for the settlement of private disputes. This is because everybody has immediate access to the apparatus of the state and uses it frequently against other members of society. . . .
>
> Since private enmity has been the primary motivation for bringing denunciations to the authorities, we are best able to understand totalitarianism's all-pervasiveness and awesome power *not* as a consequence of its being well-informed and efficient. On the contrary, the evidence suggests that it is both dismally ill-informed and mismanaged. The real power of a totalitarian state results instead from its being at the disposal, available for hire at a moment's notice, to every inhabitant.[41]

Under these conditions, everyone is a threat to everyone else, and "mutual fear and distrust" pervade society.[42] And, under these conditions, self-help and other forms of nonlegal social control tend to wither away. To bring a grievance to anyone but a government official can be dangerous, particularly if it is expressed directly to the offender, which might lead to a retaliatory complaint. Mere suspicion of being aggrieved can be dangerous. Hence, the choice is often between bringing an official complaint or doing nothing at all. Life under totalitarianism almost resembles the situation Hobbes associated with statelessness—"a war of every one against every one"—but the major weapon is law itself.[43]

Modern democracies such as the United States have also experienced ever more dependence on law. A hallmark of Western modernization, the growth of law partly reflects the differentiation—or division of labor—of society at large.[44] Functions once performed by everyone increasingly become the responsibility of specialists in everything, such as health, food, shelter, clothing, teaching, religion, transportation, communication, and recreation. Lawyers and legal officials, specialists in conflict, have evolved in the same fashion.

Modern populations are enormously dependent on specialists and sometimes helpless without them, possibly with harmful consequences. A sick or injured person might die because no physician is available, or a fire might burn out of control because there are no firefighters. When helplessness without specialists results in a worsening of problems, people are no longer merely dependent, but *over*dependent. This has happened to some extent in the handling of crime and conflict.

### The Kitty Genovese Syndrome

As people become increasingly dependent on lawyers and legal officials such as police officers and judges, their capacity to handle their own problems atrophies and occasionally disappears altogether. In modern America, this is illustrated by a phenomenon known as the Kitty Genovese syndrome.

Ms. Genovese, a young woman in New York City, died because her cries were ignored by 38 of her neighbors as she was stabbed and raped by a lone assailant near her apartment building in 1964. Many of these neighbors turned on their lights and watched from their windows, but none actively interfered. By the time the police arrived the man had run away, and Ms. Genovese was dead.

Ms. Genovese's neighbors explained afterward that they assumed someone had called the police when the screaming began, but a call was actually made only later when it was too late. In fact, the assailant glanced around continually during the attack, seemingly fearful that someone would intervene, and probably would have abandoned the attack if someone had simply rapped on a window or opened a door.[45] In any event, total reliance on the police proved to be fatal. Kitty Genovese was a victim of legal overdependency.[46]

### Law as a Cause of Crime and Conflict

Legal overdependency may increase the incidence of illegality. When citizens abdicate their responsibility for social order to legal officials, some offenses become easier to commit (because of a decline in surveillance), apprehension of wrongdoers becomes more difficult (because ordinary citizens no longer participate), and penalties may decline (because law may be more lenient than popular justice). A severe dependence on law may therefore reduce the deterrence of crime.[47] For instance, rape dramatically increased in parts of East Africa when the British introduced criminal justice and prohibited villagers from dealing with rapists and suspected rapists with their traditional (and harsher) system of self-help.[48] Other violence and crime may also increase when law replaces self-help, a pattern illustrated in African and other societies that were stateless and lawless before their conquest by Europeans: Traditionally, many of these people had a horror of violence and actively discouraged it in everyday life, but after colonial law was introduced—with specialists to handle such problems—ordinary citizens tended to relax their vigilance, and violence increased.[49] As long as law was weak on the American frontier and nearly everyone owned a gun and was ready to use it, crimes such as rape, robbery, burglary, and theft were almost unknown.[50]

We also know that conflict may be fomented and escalated by lawyers. Not only do lawyers encourage people to assert their rights, but their mere participation may alter the social structure of a conflict in such a way as to amplify it (see Chapter 1). For example, the intervention of lawyers may widen the social distance between the adversaries, such as those in a marital, employment, or commercial conflict, making an informal settlement more difficult to achieve.[51] Partly because of the dependency it engenders and partly for other reasons, then, law may worsen the conditions normally thought to make it necessary.

## Legal Minimalism

As noted earlier, the most common strategy of legal reform entails an increase of law itself. Here we explore the feasibility of precisely the opposite strategy: a reduction of law, or a delegalization of modern society. This would not only eliminate some of the discrimination so widespread in legal life, including the selective definition of illegality itself, but might also revive the capacity of ordinary citizens to handle matters now monopolized by law.

### Sociological Anarchism

Interest in delegalization is growing.[52] Even so, many regard the subject with a degree of squeamishness, perhaps because technically delegalization is a polite name for anarchism. But its sociological version differs considerably from the viewpoints usually associated with anarchism.[53] Sociological anarchism is simply a self-conscious application of sociology to the reduction of law.

Whereas anarchism traditionally involves a radical and uncompromising proposal—the abolition of the state and law—its sociological version approaches delegalization as a matter of degree. Traditional anarchism also typically involves a strong faith that society will improve, even become a "heaven on earth," when the state and law are abolished. Sociological anarchism entails no such faith or ideology, but is merely a form of social engineering concerned with creating conditions inimical to law. Finally, and surely the major reason for its bad reputation, anarchism sometimes means violent revolution. Its sociological version, however, can be gradualistic and cautiously experimental.

Anthropological evidence indicates that people have been hostile to law throughout most of human history. Hunters and gatherers, herdsmen, and those in other simple societies tend to be fiercely independent and to resist coercive authority of every kind.[54] For instance, an early observer once described the Algonkian Indians of North America as follows:

> As every one of them entertains a very high opinion of his consequence and is extremely tenacious of his liberty, all injunctions that carry with them the appearance of a command are instantly rejected with scorn. . . . There is no visible form of government; they allow of no such distinction as magistrate and subject, everyone appearing to enjoy an independence that cannot be controlled.[55]

Similarly, an African Bushman gave the following response when an anthropologist asked whether his people had headmen: "Of course we have headmen! Each one of us is headman over himself."[56] Such people inhabit a world without social conditions conducive to law and are natural anarchists.

Modern anarchists are not always revolutionary but may patiently await a spiritual awakening, or "transformation of the human heart," when law will be spontaneously abandoned.[57] Some believe that modern technology will increasingly create a decentralized way of life where law will be unnecessary—a "technoanarchy."[58] Still others choose to live among themselves without law or law-like authority.[59] Virtually all, however, have the same ideal: an intimate community of equals. Their model resembles the tribal societies that long ago disappeared from most of the world.[60]

Since law increases with such conditions as inequality, population density, social atomization, and cultural heterogeneity,[61] it would be theoretically possible to design an antilegal environment by reversing these conditions. Successful experiments in anarchism have achieved this reversal on a small scale. Replicating conditions found in stateless societies, they have unconsciously applied some of the principles of legal sociology.[62] On a larger scale, incremental changes in society, such as the equaliza-

tion and homogenization of the population, are presently counter-
acting law as well. If these trends continue, law is likely to decline
substantially more.[63] A conscious strategy of delegalization would
therefore conform to existing tendencies in the evolution of modern society.

It is possible, for example, to exploit the inverse relationship
between law and other social control discussed in the last section.
Recall its implication: the more alternatives, the less law. Increasing the availability of alternatives reduces the level of dependence on law. Such a displacement of law drains off the demand
for legal remedies with more attractive substitutes. In fact, a
movement to establish alternatives to law has already emerged in
Western societies, particularly in the United States.[64] This movement mainly encourages the creation of mediation and arbitration
systems in settings where third parties are available only through
legal action. So-called neighborhood justice centers have been
established to handle cases such as family conflicts and landlord-
tenant disputes, and similar mechanisms have been developed to
handle juvenile problems, divorce settlements, consumer complaints, and grievances in and between organizations. These inno-
vations are reducing the role of law in modern life.

Another possible tactic, more direct, simply reduces the availability of law without establishing alternatives. While the displacement of law resembles so-called maintenance programs in
the treatment of drug addiction, where a more benign drug such
as methadone substitutes for a more harmful one such as heroin,
direct delegalization resembles the treatment known as cold turkey, where a harmful drug is withdrawn entirely and the addict
forced to get along without it. Direct delegalization is not
unprecedented in modern societies. In the United States, for instance, legal prescriptions and proscriptions are often discarded,
as in the recent decriminalization of vagrancy and abortion and
the deregulation of air transportation and trucking. Legal sanctions have also been reduced. Imprisonment for debt has been
eliminated, for example, the penalties for possession of marijuana
have been lowered in some localities, and capital punishment has
declined.[65] Although a total elimination of law may be imprac-
ticable under modern conditions, its reduction to a minimum—
*legal minimalism*—is not.

## Japan as a Natural Experiment

A policy of legal minimalism implies a mistrust of law and a conscious effort to manage with as little as possible.[66] Such a policy is already followed in one of the world's most industrialized and productive nations: Japan.

Japan's extremely low rates of litigation, adjudication, and imprisonment are well known, but these rates are not—as is sometimes thought—merely a result of a peculiar aversion to law on the part of the Japanese people.[67] Rather, Japanese authorities consciously seek to minimize the use of law. This entails a combination of practices, including a restriction of the number of lawyers and judges to a level far below that dictated by the supply and demand, an unwillingness to provide judges with coercive powers in civil cases, and the establishment of mandatory conciliation procedures where litigation would otherwise occur, such as in family, landlord-tenant, and commercial disputes.[68] Japan thus provides a natural experiment in legal minimalism.

The minimization of law is a recent development in Japanese history. While the rest of the modern world experienced an enormous growth of law during the twentieth century, the Japanese did not allow this to happen. The number of lawyers and judges per capita has even declined in Japan since the 1920s. The total number of judges has not increased since 1890, so that now there is only one judge for every 60,000 persons, compared to one for every 22,000 in 1890.[69] These judges have extremely heavy caseloads—quintuple those of federal judges in the United States[70]—and the court system is almost incapacitated by overcrowding. It might be said that Japanese law is intentionally inefficient. Mandatory conciliation was also initiated in the 1920s and 1930s in various fields of law, and by 1940 nearly all civil disputes were subject to this procedure. As a result of these conditions, the Japanese have less litigation today—in absolute volume—than 50 years ago.[71]

What are the consequences of Japan's policy?[72] In the first place, because law and other kinds of social control have an inverse relationship, the stifling of the former advances the latter to a level of ever greater prominence in Japan. Legal minimalism indirectly strengthens the traditional system of paternalistic authority—probably a conscious goal of the government.[73] It also indirectly strengthens the role of reputation, or "face," as a mechanism of social control. To lose face in Japan is to lose the trust and cooperation of one's fellows and to invite ostracism—a personal and social disaster comparable to imprisonment in Western societies.

Legal minimalism seems to strengthen social cohesion as well, without which matters of face and the possibility of ostracism lose significance. Unable to rely heavily on law for protection and predictability,[74] the Japanese draw together into tightly knit groups and enforce their own standards of conduct. A system of patronage and sponsorship, or "informal suretyship,"[75] facilitates transactions between people of different groups. Anyone wishing to deal with another group must be introduced and recommended by someone already known and trusted. Commercial transactions thereby become possible without the efficient legal system commonly thought essential by Westerners:

> When sellers and lenders cannot expect to obtain relief in the event of default, either they do not sell on credit or make loans, or they take great care to ensure that they are not selling or lending to "people who default." In other words, one's reputation for trustworthiness can become a necessity of life. Furthermore, where the lesser communities and patrons provide the most important substitutes for formal sanctions, one's reputation will depend in part on affiliations and sponsorship. Thus in Japan, as most foreign businessmen and scholars know from personal experience, introductions are essential. Even law firms regularly, if politely, turn away potential clients who do not have proper introductions. Businessmen, government officials, libraries, and schools are often inaccessible without introductions.[76]

The Japanese system seems to discourage antisocial conduct more effectively than any legal system in the modern world.[77]

## Honor among Thieves

In theory, law makes trustworthiness unnecessary, even obsolete. When law is fully in command, morality itself loses relevance. Right and wrong become a specialty of professionals such as lawyers, police, and judges. Justice becomes an industry.

Even a society as legalistic as modern America, however, has some members with little or no legal protection. For example, those involved in the marketing of illegal goods and services (such as narcotics and gambling) cannot invoke the legal process when problems arise in the course of business. Their contraband is not protected as legal property, and their contracts are not enforceable in the courts. They must also guard against infiltration by the police. Their situation is thus similar to that of Japanese businessmen, only worse. Understandably, they too place a great deal of emphasis on the trustworthiness of their customers and associates. Introductions and sponsorship are crucial, and each participant strives to be regarded as a "man of honor."[78]

A minimum of law—even a complete absence—is not synonymous with chaos in modern life, then, but may actually bring about a heightened concern with trust, honor, and morality.

## A Planned Shortage of Law

An intentional reduction of law in Western societies such as the United States would increase alternative modes of conflict management such as those introduced earlier. The precise result, however, would undoubtedly differ from the response to legal shortages in Japan. For example, it might include a significant increase of the mode of conflict management most associated with lawlessness in the popular mind: self-help. And some of this would surely be violent. But a policy of legal minimalism need not apply to violence. The point of legal minimalism would be to improve the quality of life, not worsen it, and violence could receive the highest legal priority and the utmost severity.

Some of the self-help generated by a regime of legal minimalism might appreciably enhance the well-being of the population. Mutual aid between citizens would almost certainly increase,[79] and a failure to lend assistance during a crime in progress could even be defined as a crime in itself.[80] Paradoxically, therefore, law could be used to encourage self-help. A shortage of law would also encourage more self-reliance. This means that legal minimalism might partially remedy the incapacity of many citizens to manage without law. It might reduce the number of Kitty Genoveses.

But self-help is not the only alternative to law. Recall others reviewed earlier: avoidance, negotiation, informal settlement by a third party, and toleration. These too would almost certainly increase under conditions of legal minimalism. In modern America, with its already high level of mobility between regions, neighborhoods, organizations, friends, and spouses, avoidance might be a particularly attractive option.[81] So might toleration. If the availability of law were to decline, people would probably be more inclined simply to withdraw from offensive situations, or to just lump it.[82] Avoidance or toleration might take the place of police action against, say, vagrants sleeping in public or teenagers loitering on street corners. The same might apply to so-called crimes without victims such as prostitution, homosexuality, pornography, and gambling. Police work in these areas could seemingly be eliminated as surplus law without serious consequences.[83] Civil lawsuits concerning small amounts of money might be discouraged as well. At present, courts frequently serve as debt-collection agencies for businesses and professionals who grant credit to the wrong customers or clients.[84] They also serve as rent collection agencies for landlords who accept the wrong tenants. If cases such as these were eliminated, reduced, or handled in an intentionally inefficient fashion, businesses, professionals, and landlords would have to establish better means of assessing the trustworthiness of their clientele, or absorb the losses.[85] As Japan's experience indicates, a modern society can survive without much of the legal activity now regarded as natural and necessary. It might even flourish.[86]

A shortcoming commonly attributed to alternatives to law is that they may be disadvantageous for those at the bottom of society.[87] It has been asserted, for instance, that delegalization "tends to deprive those who are weaker, poorer, or of lower social status of whatever protection they had obtained from formal legal recognition of their rights."[88] Since a major goal of delegalization as contemplated here—legal minimalism—would be to lessen the degree of legal discrimination, particularly against those who are "weaker, poorer, or of lower social status," it would be ironic indeed if their plight were worsened. But this need not happen. On the contrary, legal minimalism could apply specifically to legal practices detrimental to people at the bottom of the social ladder. For instance, the depolicing of vagrants and teenagers would arguably improve their situation without creating a worse evil in the process. The same applies to the depolicing of crimes without victims. Citizens would have to develop alternatives to the police, become more tolerant, or go elsewhere. Similarly, the role of the courts in debt collection could surely be reduced without worsening the condition of the poor. New modes of conflict management would probably come into play, but it seems doubtful that the net result for all concerned would be worse.

Law could be minimized at every stage of the legal process. This might include, for example, the abolition of street harassment by the police, a practice now experienced primarily by young and unconventional individuals, often members of minority groups.[89] And, in the United States, it might include the abolition of capital punishment, a penalty reserved primarily for those who kill their social superiors.[90]

Law is notorious for giving advantages to the haves against the have-nots.[91] Some even regard this as its basic function.[92] How a shortage of law could give the haves even more advantages than they presently enjoy is difficult to imagine.[93] In sum, where legal discrimination is especially pronounced, one strategy to reduce it would be to reduce law itself. This might be a disturbing idea in modern societies with a heavy dependence on law. But perhaps it is the most obvious implication of legal sociology.

# 6

# *Conclusion*

Legal scholarship has long addressed the evolution of law over the centuries. Lawyers, historians, and social scientists have examined various dimensions of this topic, including changes in legal doctrine,[1] legal organization,[2] legal decisionmaking,[3] legal remedies,[4] and the volume and scope of legal activity.[5] But one evolutionary event has been totally ignored: the emergence of legal sociology. Not merely a matter of academic interest, this development alters the nature of law itself.

In this final chapter, we consider the relevance of legal sociology for jurisprudence and social policy. What does it say about the fundamental nature of law and justice? What new issues and challenges does it present? What solutions does it offer? The book closes with a speculation that legal sociology illustrates the growing significance of sociology in general.

First let us recapitulate: During the past decade, a new kind of legal sociology has come into being—a sociology of the case. It addresses how the social structure of cases predicts and explains their fate. We have learned, for example, that the social status of the parties, their degree of intimacy, and whether they are individuals or organizations predict and explain whether the legal process is invoked, who is likely to win, and—if the defendant loses—the severity of the remedy.

This new field of sociology entails a model of law drastically different from the jurisprudential model that portrays law as an affair of rules whose logical application decides how cases are handled. Modern jurisprudence ignores the social characteristics of the parties.[6] If a homicide occurs, for instance, who killed whom is not jurisprudentially pertinent. It does not matter whether the victim and defendant were rich or poor, white or black, one a government official and the other a vagrant, one a community leader and the other a recluse, whether they were husband and wife, friends, neighbors, or complete strangers. Such questions are, technically speaking, irrelevant: Books and articles about criminal law do not mention them. They are not mentioned when the law of homicide is taught in law school. They are not mentioned in courtroom arguments.

In the sociological model, however, such considerations are central. Who kills whom is an important predictor of how a case will be handled. For example, people who kill vagrants and prison inmates are largely immune to law. But an unemployed black ex-convict who kills a wealthy white businessman is another case altogether. Then the killer himself may even be killed by the state—a punishment rarely if ever seen when the social characteristics of the offender and victim are reversed.[7] Similarly, those who kill intimates are better off than those who kill strangers. Many are not even prosecuted.[8] The jurisprudential model treats these practices as abnormalities and condemns them as discrimination. We nevertheless find patterns of this kind in every domain of law, civil as well as criminal, throughout the world and across history. Law is socially relative.

Moreover, social differentials do not depend solely on the characteristics of the adversaries. They arise from the characteristics of other participants as well, whether third parties such as judges and jurors or partisans such as lawyers and witnesses. For example, the decisiveness and severity of judges and jurors depend on their social status and their social distance from the adversaries and their partisans. The intervention of lawyers may equalize the adversaries to some extent, and narrow or widen the social distance between them, altering the outcome of their conflict. The linguistic style of testimony is important as well, since it reflects the social characteristics of the witnesses and influences their persuasiveness. Whatever the rules and evidence, and whatever logic seems to dictate, we must also examine the social structure of a case to understand how it is handled.

Any analysis of a legal case in a social vacuum, without regard to its location and direction in social space, is incomplete and inadequate. But legal sociology is new and largely unknown in the legal establishment, and the social structure of cases is accorded little or no attention. This, however, is changing. Legal sociology is descending from its ivory tower. As this happens, the old jurisprudence taught in law schools and invoked daily by lawyers and judges loses credibility. Unsociological jurisprudence will not be forgiven indefinitely.

Modern jurisprudence distinguishes between two dimensions of law: The first is substantive and includes its content and aims. What conduct is proscribed or prescribed? What is the purpose of a particular rule? The second is procedural and specifies how rules must be formulated and applied. What determines their validity? How may a lawsuit be initiated? When is evidence admissible in court?

But cases differ socially as well as substantively and procedurally. Who has a complaint against whom? Who sides with whom? Who will decide the case? The third dimension of law is sociological.

## The Sociology of Jurisprudence

Why have legal scholars and lawyers ignored the sociological dimension of law? How could so many for so long regard law as nothing more than doctrine and logic? How can this continue in today's law schools?

Part of the answer may lie in a sociology of jurisprudence itself.

### Causes of Legal Formalism

The legal viewpoint most unlike legal sociology is that law is essentially an affair of rules.[9] This is the central tenet of the jurisprudential tradition known as legal formalism. It holds, in particular, that cases are decided by the logical application of rules to evidence. The rules—substantive and procedural—define what is subject to legal authority and how it should be handled. Legal formalism dates back at least to the Roman Empire and still dominates legal scholarship, education, and practice. Even in popular parlance, to "think like a lawyer" is to know how to apply rules logically to evidence, with special reference to how similar cases have been handled in the past.

A concern with rules, however, is not socially universal. The invocation and application of rules occurs under specifiable conditions, among which are social inequality and social distance. Parents, for example, are more likely to invoke rules against their children than against each other, and the same applies to employers and employees, teachers and students, masters and slaves, and the like. Third parties socially superior to the adversaries are more likely to invoke rules than those closer in social status.[10] People in intimate relationships, such as married couples and friends, are also relatively unlikely to resolve their differences by the logical application of rules.[11] But an invocation of rules becomes more likely when a third party socially distant from the adversaries intervenes to handle their dispute.[12] Both equality and intimacy thus tend to inhibit the use of rules.

We can observe these tendencies across societies as well as across cases. Stratified and impersonal societies are more rule-oriented than egalitarian and intimate societies. Explicit rules are almost unknown in simple tribes of hunters and gatherers, where differences in wealth and other forms of social status are small and everyone is closely acquainted. They might have a few rules about sex and marriage or sharing food, but that is all.[13] In fact, when anthropologists ask tribal people to describe their "laws" or "customs" they often encounter a total lack of understanding. Some tribes have no conception of rules at all,[14] and others only vague expectations.[15] Rules become more prominent as societies grow, differentiate, and lose their personal character. Imperial Rome, for instance, was more rule-oriented than Republican Rome.[16] Codes of law—lists of rules—were first compiled in societies with a high degree of inequality and social distance among and between their inhabitants: Babylonia, classical India, ancient Israel, the early Germanic kingdoms, etc. Legal formalism, the rule-oriented conception of law, will therefore prevail only where social conditions conducive to rules are found. Those conditions—inequality and social distance—occur in abundance in modern societies such as the United States.

But even if judges and lawyers continually invoke legal rules in courtrooms, the rules alone do not predict and explain how cases are handled. Like cases are not always treated in like manner. Technically similar cases will be handled in a similar fashion only when they are—or appear to be—socially similar as well. Only then does the credibility of legal formalism remain intact. There is reason to believe, however, that the social similarity of cases has been decreasing, at least in modern America.

Since the turn of the twentieth century, if not earlier, legal life in the United States has been democratizing and diversifying. This is most noticeable in the administration of law. An ever broader spectrum of the population participates as police officers, prosecutors, judges, and jurors. Judges, for example, have traditionally been prosperous white men, middle-aged or older, ethnically and religiously conventional. But this is changing. The judiciary increasingly includes people from many walks of life—those of modest wealth, younger people, women, blacks, Hispanics, Asian-Americans, Catholics, Jews, and so on. Third parties of all kinds are more socially heterogeneous than ever before. So are lawyers. This diversification increases the degree of variation in how cases are processed and decided, and so increasingly the success of litigants depends on who handles their cases. In turn, the credibility of legal formalism weakens: If the handling of cases varies with the social characteristics of judges and juries, what is the role of the written law? Even lawyers are becoming skeptical about the relevance of the rules. Many compare justice to a slot machine, as if the handling of cases were as unpredictable as a game of chance. But they are wrong. Justice is unpredictable only when the predictions are based on the written law alone.

American litigants are diversifying as well. For example, prosecutors still bring criminal cases primarily against the poor and the marginal, but increasingly those of higher status find themselves in jeopardy as well.[17] We see ever more prosecutions of so-called white-collar offenses, even against respected politicians, executives, and corporations. New prohibitions, such as consumer- and environmental-protection laws, apply disproportionately to the comfortable classes. New users of law also diversify the cases.[18] Those at the bottom of society who once avoided legal institutions increasingly rely on the police and other officials. Legal-assistance programs make it easier to initiate civil cases, as do small-claims courts. A multitude of regulatory agencies, such as housing and antidiscrimination commissions, broaden legal participation still further.[19] At the same time, the jurisdiction of law is expanding substantially, opening the courtroom to ever more sectors of the population.[20] Although the role of law is often exaggerated in modern America,[21] it has never been greater.

As the litigants diversify, so do legal outcomes, adding to the variation arising from the diversification of third parties. It is ever more difficult to predict how cases will be handled with the written law alone. The significance of the rules is declining. Rules proliferate when people are socially unequal and atomized, and these conditions still exist. But rules predict the handling of cases only when technically similar cases are socially similar as well, and these conditions are disappearing. Legal formalism is losing its social foundations.

## Causes of Legal Sociology

The declining significance of rules has important implications for modern jurisprudence. If the rules do not predict and explain the handling of cases, what does?

Enter sociology.

Legal sociology offers a new understanding of legal life by studying what jurisprudence ignores: the social structure of the cases. It begins where legal education ends, discovering order where legal formalism finds chaos. Legal sociology documents the significance of social diversity and specifies its impact on the handling of cases. Moreover, it arises and prospers under the same conditions that undermine legal formalism: the increasing diversification of the cases. It demonstrates the social relativity of law.

Legal sociology also reflects a new location in the social structure of legal scholarship. It is largely the creation of nonlawyers, newcomers to the study of law and outsiders to the legal establishment. Socially removed from most lawyers, judges, and law

professors, sociologists embrace a different model of law empha-
sizing different aspects of legal reality.[22]

Legal knowledge is socially determined.[23]

## The Jurisprudence of Sociology

As outlined in Chapter 2, legal sociology can readily be applied
to the practice of law. Sociological litigation appears inevitable,
if only because it allows lawyers to pursue their own interests
more effectively. If lawyers know the relevance of the social
characteristics of cases, they can apply this knowledge to such
matters as the screening of clients, the decision whether to settle
or go to trial, the preparation of cases, the selection of witnesses,
and the choice of a judge or jury. Legal sociology may increasingly become part of legal education. Law students may learn
to assess the sociological merits of a case as well as the technical
merits traditionally taught. They may learn that a technically
strong case can be sociologically weak, and vice versa, and that
this may be critical for a client. For that matter, the clients themselves may increasingly expect lawyers to know legal sociology.
And if ordinary citizens become sociologically sophisticated, this
will influence their own legal behavior, including their choice of
lawyers.

### Unintended Consequences of Legal Sociology

The advent of sociological litigation, however, is not likely to be
universally acclaimed. Some will surely regard it as morally re-
pugnant. That social differentials pervade law—the central finding of legal sociology—is disturbing enough. It implies, after all,
a gross violation of the rule of law. Law is supposed to be au-
tonomous, not a dependent variable predictable from the social
structure of cases.[24]

But sociological litigation goes further: It takes social differentials for granted. It accepts discrimination as a starting point,
the first principle of effective legal action. Not merely a loss of
legal innocence, this must seem a new form of prostitution and a
perversion of legality itself. Discrimination would become a professional service. An art.

Sociological litigation would even increase discrimination. If, for example, lawyers consistently apply the principle that downward cases (against social inferiors) are more likely to succeed than upward cases, they will engage in practices such as the following: They will become increasingly reluctant to represent people with grievances against social superiors, or demand higher fees for doing so; they will more readily advise defendants in downward cases to plead guilty or otherwise capitulate; they will appeal more cases to higher courts when adverse decisions contradict this principle (i.e., when their downward cases fail or when upward cases succeed against them). Similarly, lawyers applying the principle that the success of legal cases varies directly with relational distance will more willingly represent people with grievances against strangers than those with grievances against their family members, friends, or other intimates; they will increasingly advise defendants to capitulate to strangers but to resist intimates; they will appeal more unsuccessful cases against strangers and unsuccessful defenses against intimates of their clients.

If these and other practices described in Chapter 2 occur, social differentials in legal life will increase. People with grievances against social superiors will have less help from law, and those brought to court by social superiors will be urged to give up. Since social inferiors will be less able to appeal if they lose, reversals of their cases will decline. Those with grievances against intimates will become less litigious and those brought to court by strangers less likely to defend themselves. Wealth, organization, conventionality, and respectability will become more fateful, along with every other factor identified as legally relevant by sociologists. The probabilities will grow more certain. Legal sociology will become a self-fulfilling prophecy.[25]

## Law as a Social Problem

Legal sociology is not merely an academic exercise, a new branch of legal scholarship and a revision of traditional jurisprudence. It poses a crisis for law itself. Its central finding that the handling of cases is socially relative—that discrimination is ubiquitous—devastates any claim that the rule of law prevails, that like cases are treated in like fashion. And when legal sociology enters the practice of law, its predictions become a guide to legal action, routinizing and marketing social differentials. Discrimination sharpens and hardens. Law gets worse.

But we can imagine other possibilities. Because legal sociology increases the visibility of discrimination in legal life, it may awaken public indignation. The situation may seem intolerable, a social problem that must be solved. This in turn may provoke legal reforms to reduce discrimination. Although large-scale measures such as a redistribution of wealth are unlikely in the near future, more limited reforms such as those presented in earlier chapters might now become attractive. One of these would homogenize the social structure of legal cases by assigning them to organizations (legal co-ops). This would counteract the present disadvantages of individuals and equalize cases in other respects. A second reform would homogenize the cases by reducing information about their social characteristics (desocialization). This would correspondingly reduce the legal impact of social characteristics as well. A third reform would minimize the jurisdiction and application of law itself (delegalization). Some of the social differentials in the definition and handling of illegality would disappear, and alternative modes of conflict management would increase, weakening the widespread dependence on legal officials and possibly strengthening the role of trust and morality in modern society.

The findings of legal sociology are socially unacceptable because one proposition is so firmly established in modern jurisprudence: Discrimination is wrong. Virtually every legal thinker takes for granted that every trace of discrimination should be banished. Nothing seems more obvious. But let us examine this proposition more closely. Perhaps it is unrealistic.

## A New Morality of Law

Although basic to modern jurisprudence, the principle that like cases should be treated in like fashion is not universal across societies. On the contrary, it is unusual. Most people throughout human history would find it difficult to understand, even peculiar.

## Modern Jurisprudence as a Scientific Curiosity

People everywhere have nearly always regarded the social structure of conflict as fundamental to its resolution. They have nearly always believed it right and proper to take the social characteristics of each case into account. They have done this openly, without embarrassment. They would have difficulty understanding how anyone could handle a case without considering whether the parties are closely related, whether they live in the same household, whether they are social equals, the history of their relationship, the allies of each, whether they are male or female, old or young, and so on. Without information about the social structure of a case, people in ancient and tribal societies would not know what to do. Their jurisprudence has always been explicitly sociological.

For example, the Code of Hammurabi, compiled about 2000 B.C. in Babylonia, specifies a different remedy for each social configuration in which an offense might occur:

> If a man strike another man of his own rank, he shall pay one *mana* of silver.
> If a man strike the person of a man . . . who is his superior, he shall receive sixty strokes with an ox-tail whip in public. . . .
> If a man's slave strike a man's son, they shall cut off his ear.[26]

All ancient civilizations have qualified their written law sociologically.[27] The same applies to tribal societies, though for them social considerations generally are a matter of custom rather than written law.[28] The Nuer of Sudan, for instance, explicitly vary their response to homicide with the degree of intimacy between the killer and the victim:

> The killing of a stranger, especially of a foreigner, who does not come within the most extended form of the social structure, is not really a wrong (*duer*) at all. . . .
> The killing of a fellow tribesman, and to a much lesser extent a fellow Nuer of another tribe who is within the orbit of the killer's social sphere, is a wrong because it is an offense against the stability of society in its most extended form. But it is a private

delict and not a crime, and demands only retaliation or restitution. The closer the relationship between the component tribal
segments involved, the greater the sanction for restitution. . . .

Finally, in the narrowest definition of blood-relationship, where
kinship is a reality and not merely a fictional social form—that is,
within the lineage or the extended family group—restitution becomes less and less necessary because the persons who assist in
the payment of compensation are also the recipients. [Thus,] a
man does not pay compensation at all if he kills his own wife—a
rare occurrence—for he would have to pay it to himself.[29]

The primary difference between ancient civilizations and tribal
societies in this respect is that the former were aristocracies—
usually with slaves—and tended to emphasize status differences,
whereas the latter are more egalitarian and tend to emphasize relational differences.

Social characteristics of the parties were sometimes explicitly
addressed in medieval European law,[30] but over the centuries this
practice largely disappeared. Such factors as the age and criminal record of defendants may still be regarded as relevant, but
the elimination of social considerations from the written law of
Europe and America continues. In the United States, for example, the immunity of spouses from rape charges and negligence
actions has been challenged in court and seems unlikely to survive. The state has enjoyed certain immunities as well, but these
too are narrowing.[31] The evolutionary pattern is clear: The modernization of society entails an increasing reluctance to qualify
legal rights and liabilities with social considerations such as
wealth, race, ancestry, occupation, and the nature of the relationship between the parties.[32] Increasingly, law is solely concerned with "subjecting human conduct to the governance of
rules."[33] Who engages in prohibited conduct (or fails to engage
in prescribed conduct), who complains about it, and who handles it, are not addressed. More precisely, these matters are irrelevant. This is theory, however, not fact.

### The Relevance of Reality

Even if the legal doctrines of modern societies do not explicitly
address the social characteristics of cases, this does not mean

those characteristics are extraneous to the handling of cases. Social differentials are not peculiar to ancient or tribal law. Rather,
the primary difference between early and modern law is that today's legislatures and courts do not officially acknowledge these
differentials. This may have been understandable before the rise
of legal sociology, but no longer. The findings and formulations
reviewed in this book are now readily available and steadily
growing. Legal sociology therefore invites modern jurisprudence
to face reality: Cases are not decided by rules alone. Every case
has social characteristics that predict how it will be handled.
Perhaps legal thinkers should reconsider the proper relevance of
these characteristics.

A new sociological jurisprudence[34] would critically examine
each of the social factors associated with the handling of cases:
the several kinds of social status, the involvement of organizations (including the state), the degree of relational distance between the parties, the availability of alternatives, and the like.[35]
Which of these, if any, should be legally endorsed? Which, if
any, should be counteracted?

Consider, for example, the proper relevance of relational distance. We know that the degree of intimacy between the adversaries is a powerful predictor of how a case will be handled: The
more intimacy, the less law. A man who beats or kills his wife or
friend is likely to be handled more leniently than a man who
commits the same offense against a stranger. The evidence is
overwhelming (see Chapter 1). The same applies to all crime
and to breaches of contract and torts. This is relational discrimination, and it cries out for recognition. But is it right or wrong?

How law should behave lies beyond the jurisdiction of sociology. Moreover, the proper relevance of the social structure of a
case is not obvious. Some, for instance, might prefer leniency for
those who victimize friends and relatives. We know that people
are less likely to report such crimes to the police, but are they
less blameworthy? People are also less likely to bring lawsuits
for negligence or breach of contract against intimates, but should
there be a formal reduction of liability in such cases? Some
might find it morally worse for, say, a brother to victimize his
sister than for a stranger to do so, but might nevertheless feel the
brother should be handled less severely in court. Law, they might

argue, is not an appropriate remedy for intimates. Some might regard those who victimize strangers as a greater danger to the community,[36] but others might feel that people need at least as much protection from intimates as from strangers, or that it is even worse—more terrifying and degrading—to be victimized by an intimate at home than by a stranger on the street. Still others might argue that none of these considerations is legitimate, that we should hold everyone to the same standard. The point here is that the issue is largely unexplored.

We also know that organizations enjoy legal advantages (see Chapter 3). Should they? And those who victimize their social superiors—the wealthier, the more integrated, the more conventional—receive harsher treatment. Is this right? Is a wealthy white's life or limb worth more than a poor black's? A socialite's more than a vagrant's? Should it matter whether alternatives to law are available to handle a particular case? Should a defendant's past record be taken into account? Perhaps a criminal record, but a school, employment, or military record? Should unemployed and unmarried people have disadvantages?

A moral inquiry into these questions might well conclude that some existing practices are acceptable or even desirable. These might include the advantages of people who victimize their intimates, those for whom alternatives to law are available, and those without a criminal record. The advantages of social status, however, might be deemed unacceptable. Some practices might even be condemned to destruction by any means necessary. For example, American courts continue to apply capital punishment disproportionately to blacks who victimize whites. All efforts to change this pattern have failed.[37] Perhaps, therefore, an appropriate solution would be to eliminate capital punishment itself.

A new sociological jurisprudence would acknowledge that a conception of law as an affair of rules alone is incomplete and obsolete. It would resemble the jurisprudence of ancients and primitives, for whom social considerations are central. Yet it would also introduce a new stage in the evolution of legal thought.

## The Age of Sociology[38]

Sociological knowledge has applications in the practice of law, in legal reform, and in jurisprudence and social policy. No aspect of legal life is immune. Moreover, the applicability of legal sociology illustrates the growing significance of sociology in general. Future sociological inquiries seem likely to yield practical applications and policy implications as unexpected and unsettling as those found in legal sociology.

Consider, for example, the field of conflict management. As noted in Chapter 5, law belongs to a vast universe of practices by which people handle grievances. This universe includes diverse forms of self-help, avoidance, negotiation, settlement, and toleration. Law is rare. But sociological research reveals that all these possibilities—from violence to gossip—resemble law in one respect: They are socially specific.[39] How anyone judges anyone else, and what remedy seems appropriate, depends on whether they are rich or poor, white or black, male or female, adult or child, corporate or individual. Morality is socially relative, and so is ethics, honor, etiquette—every conception of what is good and evil, right and wrong, proper and improper. The same applies to evaluations of everything else, to what is beautiful or ugly, important or trivial, true or false.[40] Sociology changes our conception of judgment itself.

Sociology also has endless implications for the nature of social justice. Intimacy, for example, which tends to immunize people against law, has numerous consequences in other realms of social life as well. Individuals with "contacts" have advantages getting a job,[41] scientists in social networks have advantages getting resources and recognition,[42] and people in distress even slightly acquainted with a bystander have advantages getting help.[43] Class, race, and ethnic discrimination have long been recognized, but the advantages of intimacy do not receive similar attention. Sociology makes relational discrimination visible, a possible frontier of modern morality. It raises relational consciousness, and consciousness of everything social.

We cannot assess the ultimate significance of sociology. Social space is still being discovered and explored. Beyond social consciousness of any kind—class, race, or relational—lies sociological consciousness. Sociological knowledge will surely grow, creeping throughout the world with a fatefulness unknowable. We only know that the science of society changes society. Social action increasingly becomes a self-conscious application of sociology. Sociological power is harnessed, a new form of energy generated, its uses impossible to imagine. A sociological society.

This must be a new stage of human evolution.

# Notes

See References for complete bibliographic information about the items cited below.

## Chapter 1

1. By the present author; reprinted in Black and Mileski, *Social Organization of Law.*

2. See, e.g., Skolnick, *Justice without Trial;* Mayhew, *Law and Equal Opportunity;* Lefstein et al., "In search of juvenile justice."

3. For a recent discussion of these issues, see Cooney, "Behavioural sociology of law: a defence."

4. See, e.g., Maine, *Ancient Law;* Durkheim, *Division of Labor;* Weber, *Law in Economy and Society.*

5. See, e.g., Cain and Hunt, *Marx and Engels on Law;* Fitzpatrick, *Law and the State in Papua New Guinea;* Snyder, *Capitalism and Legal Change;* Beirne and Quinney, *Marxism and Law.*

Marxian scholarship on law is typically concerned with the question of whether legal policy has a class bias. The usual conclusion is that such a bias does exist, in that legislation and other legal policies tend ultimately to serve the interests of economic elites. But see Beirne, "Empiricism and the critique of Marxism on law and crime"; "Some more empiricism in the study of law," 473.

Among lawyers, the Marxian tradition is partially followed by the proponents of "critical legal studies," a scholarly movement that regards law as an ideology supporting the *status quo.* See, e.g., Kairys,

*Politics of Law;* Kennedy, *Legal Education and the Reproduction of Hierarchy,* Chapter 2; *Stanford Law Review* Editors, "Critical legal studies."

6. See, e.g., Holmes, "Path of the law"; Pound, "Law in books and law in action"; Rodell, *Woe unto You, Lawyers!;* Frank, *Courts on Trial;* Llewellyn, *Jurisprudence;* Twining, *Karl Llewellyn and the Realist Movement.*

7. The critical legal studies movement, mentioned above, also expresses skepticism about the centrality of rules and principles in legal decisionmaking. It can thus partly be construed as a new version of legal realism. See Livingston, " 'Round and 'round the bramble bush."

8. Llewellyn and Hoebel, *Cheyenne Way.*

9. Gluckman, *Judicial Process among the Barotse of Northern Rhodesia.*

10. E.g., Bohannan, *Justice and Judgment among the Tiv;* Gibbs, "Kpelle moot"; Gulliver, *Social Control in an African Society;* Nader, "Analysis of Zapotec law cases"; Fallers, *Law without Precedent;* Collier, *Law and Social Change in Zinacantan;* Koch, *War and Peace in Jalémó.*

11. E.g., LaFave, *Arrest;* Skolnick, *Justice without Trial;* Black, "Production of crime rates"; "Social organization of arrest"; Reiss, *Police and the Public;* Smith and Klein, "Police control of interpersonal disputes."

12. E.g., Newman, "Pleading guilty for consideration"; *Conviction;* Skolnick, "Social control in the adversary system"; Buckle and Buckle, *Bargaining for Justice;* Mather, *Plea Bargaining or Trial?;* Maynard, *Inside Plea Bargaining.*

13. E.g., Mileski, "Courtroom encounters"; Vera Institute, *Felony Arrests;* Boris, "Stereotypes and dispositions for criminal homicide"; Feeley, *The Process Is the Punishment;* Bowers and Pierce, "Arbitrariness and discrimination under post-*Furman* capital statutes"; Kruttschnitt, "Women, crime, and dependency."

14. E.g., Macaulay, "Non-contractual relations in business"; Mayhew, *Law and Equal Opportunity;* Mayhew and Reiss, "Social organization of legal contacts"; Ross, *Settled Out of Court;* Mileski, Policing Slum Landlords; Rosenthal, *Lawyer and Client;* Wanner, "Public ordering of private relations," Parts 1 and 2; Yngvesson and Hennessey, "Small claims, complex disputes"; Trubek et al., *Civil Litigation Research Project,* Silberman, *Civil Justice Process;* Wheeler et al., "Do the 'haves' come out ahead?"

15. Merry, "Going to court."

16. Baumgartner, "Social control in suburbia"; Law and the middle class"; *Moral Order of a Suburb.*

17. Engel, "Oven bird's song."

18. Lundsgaarde, *Murder in Space City.*

19. Thomas-Buckle and Buckle, "Doing unto others."

20. Ellickson, "Of Coase and cattle."

21. Steele, "Fraud, dispute, and the consumer"; Macaulay, "Lawyers and consumer protection laws"; Nader, *No Access to Law.*

22. Banton, *Policeman in the Community;* Manning, *Police Work.*

23. Punch, *Policing the Inner City.*

24. Kalogeropoulos and Rivière, "Police station discourse."

25. Yngvesson, "Responses to grievance behavior"; "Atlantic fishermen."

26. Todd, "Litigious marginals."

27. Starr, *Dispute and Settlement in Rural Turkey;* "Turkish village disputing behavior."

28. Ayoub, "Conflict resolution and social reorganization in a Lebanese village"; Rothenberger, "Social dynamics of dispute settlement in a Sunni Muslim village in Lebanon"; Witty, *Mediation and Society.*

29. Berman, "Cuban popular tribunals."

30. Kawashima, "Dispute resolution in contemporary Japan"; Bayley, *Forces of Order;* Haley, "Myth of the reluctant litigant"; "Sheathing the sword of justice in Japan"; Ames, *Police and Community in Japan;* Miyazawa, "Taking Kawashima seriously."

31. Garnsey, "Legal privilege in the Roman Empire"; Lintott, *Violence in Republican Rome.*

32. Offner, *Law and Politics in Aztec Texcoco.*

33. Ruggiero, *Violence in Early Renaissance Venice.*

34. Kagan, *Lawsuits and Litigants in Castile.*

35. van der Sprenkel, *Legal Institutions in Manchu China.*

36. Samaha, *Law and Order in Historical Perspective;* Hay et al., *Albion's Fatal Tree;* Hanawalt, *Crime and Conflict in English Communities;* Beattie, *Crime and the Courts in England.*

37. Baumgartner, "Law and social status in colonial New Haven"; Hindus, *Prison and Plantation;* Friedman and Percival, *The Roots of Justice;* Nelson, *Dispute and Conflict Resolution in Plymouth County;* Ayers, *Vengeance and Justice;* McGrath, *Gunfighters, Highwaymen, and Vigilantes.*

38. Personal communication from a former employee of a Midwestern city morgue.

39. Lundsgaarde, *Murder in Space City,* especially 232.

40. See, e.g., Roberts, "The unwritten law."

41. See, respectively, Biderman, "Surveys of population samples for estimating crime incidence"; Ross, *Settled Out of Court;* Macaulay, "Non-contractual relations in business"; Nader, *No Access to Law;*

Black, *Manners and Customs of the Police;* Trubek et al., *Civil Litigation Research Project.*

42. Trubek et al., *idem,* S-19.

43. *Idem.*

44. *Idem,* S-23.

45. Calculated from *idem,* S-19 and S-23.

46. See Black, *Behavior of Law,* 3; "Note on the measurement of law."

47. See Black, *Behavior of Law,* Chapters 2–6.

48. It should be noted that insofar as race is a biological variable it has no inherent sociological relevance. Even so, in Western societies such as modern America, blacks in the aggregate have significantly less income, education, and other resources. Race therefore serves as a crude indicator of social status. Compare Fields, "Ideology and race in American history."

The legal relevance of race is treated more fully in Chapter 4.

49. Calculated from Bowers and Pierce, "Arbitrariness and discrimination," 594.

50. *Idem.*

51. For other examples, see Black, *Behavior of Law,* 17–20.

52. See *idem,* Chapters 2–4 and 6.

The above formulation and those in the following pages apply to legal behavior only when other variables known to be relevant—and specified in the same body of theory—are constant. Besides social status, for example, the degree of intimacy between the parties and whether they are individuals or organizations are relevant variables and must remain constant if the effect of social status is to be predicted with confidence. The author's *Behavior of Law* provides an overview of these variables and their theoretical relevance.

A considerable literature addresses the strengths and weaknesses of this body of theory. See, e.g., Baumgartner, "Law and social status in colonial New Haven"; "Law and the middle class"; Gottfredson and Hindelang, "A study of *The Behavior of Law*"; Merry, "Going to court"; Myers, "Predicting the behavior of law"; Kruttschnitt, "Social status and sentences of female offenders"; Vago, *Law and Society,* 48–50; Greenberg, "Donald Black's sociology of law"; Horwitz, "Resistance to innovation in the sociology of law"; Hunt, "Behavioural sociology of law: a critique of Donald Black"; Cooney, "Behavioural sociology of law: a defence"; Griffiths, "Division of labor in social control"; Dannefer, " 'Who signs the complaint?' "; Staples, "Toward a structural perspective on gender bias in the juvenile court"; Silberman, *Civil Justice Process,* Chapter 5; Tomasic, *Sociology of Law,* 24–29; Hembroff, "Seriousness of acts and social contexts."

53. Calculated from Bowers and Pierce, "Arbitrariness and discrimination," 594. American blacks accused of raping a black are extremely difficult to convict, much less punish. But when the alleged victim is white, the conviction rate increases substantially. Holmstrom and Burgess, *Victim of Rape*, 144 and 249.

54. Landy and Aronson, "Influence of the character of the criminal and his victim on the decisions of simulated jurors," 150.

55. Bowers and Pierce, "Arbitrariness and discrimination," 594.

56. Black, *Behavior of Law*, Chapters 2–4 and 6. For some specific instances, see Gough, "Caste in a Tanjore village," 48; van der Sprenkel, *Legal Institutions in Manchu China*, 27; Garnsey, "Legal privilege in the Roman Empire"; Welsh, "Capital punishment in South Africa," 416; Inglis, '*Not a White Woman Safe*', Chapter 3; Baumgartner, "Law and social status in colonial New Haven."

57. See, e.g., Garfinkel, "Research note on inter- and intra-racial homicides"; Bowers and Pierce, "Arbitrariness and discrimination."

58. See, e.g., Johnson, "The Negro and crime"; Myrdal, *American Dilemma*, 550–555; Rubin, "Disparity and equality of sentences," 65–67.

59. Relational distance refers specifically to "the degree to which [people] participate in one another's lives" and can be measured by "the scope, frequency, and length of interaction between people, the age of their relationship, and the nature and number of links between them in a social network." Black, *Behavior of Law*, 40–41.

60. See, e.g., McIntyre, "Public attitudes toward crime and law enforcement," 45; Block, "Why notify the police," 560–561.

61. Black, "Production of crime rates," 740–741.

62. Black, "Social organization of arrest," 1097–1098; *Manners and Customs of the Police*, Chapter 5.

63. Hall, *Theft, Law, and Society*, 318; see also Lundsgaarde, *Murder in Space City*, 232.

64. Vera Institute, *Felony Arrests*, 23–42. This report shows that defendants having a prior relationship with their alleged victim are less likely to be convicted. But assault cases occurring between people in a prior relationship that do result in a conviction are more likely to involve a weapon and serious injury than cases between strangers. For this reason, the report does not provide meaningful evidence on the matter of differential penalties. *Idem*, 29–31. We know, however, that *fatal* assaults between intimates tend to result in less severe penalties. Lundsgaarde, *Murder in Space City*, 232; Gross and Mauro, "Patterns of death," 58–59.

65. Black, "Social control of the self."

66. E.g., Turnbull, *Wayward Servants,* 140 and 274–275; Reid, *Law of Blood,* 117–118; Lee, *!Kung San,* 452–453.

67. See, e.g., Black, "Production of crime rates," 740–741; "Social organization of arrest," 1097–1098; *Manners and Customs of the Police,* Chapter 5; Lundsgaarde, *Murder in Space City;* Vera Institute, *Felony Arrests;* Merry, "Going to court"; Holmstrom and Burgess, *Victim of Rape,* 246–247; Williams, "Classic rape," 463.

68. Gross and Mauro, "Patterns of death," 58–59.

69. Macaulay, "Non-contractual relations in business."

70. Engel, "Oven bird's song."

71. See, e.g., Bohannan, *Justice and Judgment among the Tiv,* 210; Gulliver, *Social Control in an African Society,* 204; Kawashima, "Dispute resolution in contemporary Japan," 43–45; Tanner, "Selective use of legal systems in East Africa," 7; Engel, *Code and Custom in a Thai Provincial Court,* 143; Kagan, *Lawsuits and Litigants,* 18–20.

72. Black, *Behavior of Law,* 40–46. More precisely, the relationship between law and relational distance is curvilinear, since people from completely different worlds (such as different tribes or societies) are, like intimates, comparatively unlikely to resort to law when a conflict arises. Within a society such as modern America, however, the relationship seems to be approximately linear: the more relational distance, the more law.

73. See generally Black, *Behavior of Law.*

74. Although variables such as the social status of the adversaries and their degree of intimacy predict differentials in legal behavior, no claim is being made here that anyone *intentionally* discriminates against anyone else in the legal process, or that these people even know how their own behavior compares with that of others faced with similar decisions. When a black rapes a white, for example, a judge or jury may respond to this as more serious—more traumatic for the victim—than when a black rapes another black, but race itself may not consciously influence this emotional response. Each case is handled in isolation from others, allowing emotion to operate in each one without regard to the rest. Social differentials may thereby accumulate unwittingly, without intent or a plan of any kind. This is one reason it is difficult to reduce these differentials, even when they are publicly acknowledged and conscious efforts to eliminate them are made. See, e.g., Bowers and Pierce, "Arbitrariness and discrimination." For this reason, moreover, an increased awareness of the findings of legal sociology may not result in conscious efforts by legal actors to behave otherwise.

75. See, e.g., Ladinsky, "Careers of lawyers, law practice, and legal institutions"; Heinz and Laumann, *Chicago Lawyers.*

76. See, e.g., Miller and Schwartz, "County lunacy commission hearings"; Mileski, "Courtroom encounters," 486–492; compare Feeley, *The Process Is the Punishment,* 144–145.

77. Griffiths, "What do Dutch lawyers actually do in divorce cases?," 149.

78. See Black and Baumgartner, "Toward a theory of the third party."

79. *Idem,* 113; see also Black, "Social control as a dependent variable," 24.

80. Uhlman, *Racial Justice,* 65–69.

81. Levin, "Urban politics and judicial behavior."

82. Nagel, "Judicial backgrounds and criminal cases."

83. See, e.g., Frank, *Law and the Modern Mind,* Chapter 16; Kalvan and Zeisel, *American Jury,* Chapter 1; Rosenthal, *Lawyer and Client,* 91–92; Wishman, *Confessions of a Criminal Lawyer,* 71.

84. See Kalven and Zeisel, *American Jury,* Chapters 5–6.

85. Bernard, "Interaction between the race of the defendant and that of jurors in determining verdicts," 109.

86. Hastie et al., *Inside the Jury,* 135–142.

87. Strodtbeck et al., "Social status in jury deliberations."

88. Bernard, "Interaction between the race of the defendant and that of jurors," 109–110.

89. See Black, "Social control as a dependent variable," 21–22.

90. See Simmel, *Sociology of Georg Simmel,* 149–153.

91. Engel, "Oven bird's song."

92. See Wishman, *Confessions of a Criminal Lawyer,* 74–75.

93. See, e.g., Chevigny, *Police Power;* Black, *Manners and Customs of the Police,* 174.

94. Black and Baumgartner, "Toward a theory of the third party," 113; Black, "Social control as a dependent variable," 21–23.

95. Black and Baumgartner, *idem;* see also Black, *idem,* 22–23.

96. See Black, *idem,* 24–26, for illustrations.

97. Black and Baumgartner, "Toward a theory of the third party," 114.

98. See generally O'Barr, *Linguistic Evidence.*

99. *Idem,* 72.

100. *Idem,* 74.

101. *Idem,* 80.

102. *Idem,* 90–91.

103. See, e.g., Hart, *Concept of Law;* Fuller, *Morality of Law.* Here and elsewhere in the text, the concept of "rules" refers generally to the doctrines of law that prescribe, proscribe, and permit vari-

ous modes of conduct. This includes broad principles of law such as the right to "due process" and "equal protection" guaranteed by the American Constitution. See, e.g., Dworkin, *Taking Rights Seriously.*

104. Legal officials may disregard the written law, however. See, e.g., Black, *Manners and Customs of the Police,* 180–186. Moreover, legal doctrines and the facts are often ambiguous, with uncertain implications.

105. See generally Black, *Behavior of Law.*

## Chapter 2

1. Here litigation refers broadly to the handling of cases at all stages of the legal process.

2. See generally Black, *Behavior of Law,* Chapters 2–5.

3. See, e.g., Mayhew and Reiss, "Social organization of legal contacts"; Wanner, "Public ordering of private relations," Parts 1 and 2; Silberman, *Civil Justice Process.*

4. See Black, "Compensation and the social structure of misfortune," 576–581.

5. Wanner, "Public ordering of private relations," Part 2, 302.

6. *Idem,* 300.

7. *Idem.*

8. In the United States, the examination of prospective jurors in court—known as *voir dire*—is sometimes conducted by the judge and sometimes by the attorneys. In both cases attorneys participate in the selection process by exercising their right to remove particular individuals from consideration. They may do this by arguing that an individual is prejudiced against their side ("challenge with cause") or, in a limited number of instances, by requesting an individual's removal without giving grounds ("peremptory challenge").

9. Feeley, *The Process Is the Punishment,* 173.

10. *Idem.*

11. O'Barr, *Linguistic Evidence,* Chapters 4–5.

12. *Idem,* 93–94.

13. Visual exposure and eye contact might seem to increase intimacy only trivially, but experimental evidence suggests otherwise. For example, people are more likely to lend assistance to a stranger in distress if eye contact has previously occurred. Ellsworth and Langer, "Staring and approach." Mere visual exposure of one person to another has the same effect. Piliavin et al., "Good Samaritanism."

14. Sociological litigation would not only apply knowledge of social

differentials to the practice of law. To an extent, it would perpetuate and sharpen these differentials, making legal sociology into a self-fulfilling prophesy. This possibility is discussed in Chapter 6.

## Chapter 3

1. Maybury-Lewis, *Akwē-Shavante Society*, 179. A woman can make a legal complaint only if a man sponsors it as his own, which happens very infrequently. This may help explain why Shavante women are more violent than men:

> If two women become involved in a dispute which does not concern the men, there is no forum in which it can be discussed and no institutionalized procedure for its resolution. As a result women occasionally come to blows, a thing men never do (*idem*).

2. See, e.g., Hasluck, *Unwritten Law in Albania;* Howell, *Manual of Nuer Law;* Gulliver, *Social Control in an African Society;* Peters, "Some structural aspects of the feud among the camel-herding Bedouin of Cyrenaica"; Reid, *Law of Blood;* Koch, *War and Peace in Jalémó;* Cooney, Social Control of Homicide.

Tribal societies may also be more individualistic in their management of conflict. This applies, for instance, to many hunters and gatherers, such as the Yurok of California (Kroeber, "Law of the Yurok Indians"), the Indians of the Great Plains such as the Comanche and Kiowa (Hoebel, *Law of Primitive Man*, Chapter 7), the Bushmen of Botswana (Lee, *!Kung San*), the Hadza of Tanzania (Woodburn, "Minimal politics"), and the Pygmies of Zaire (Turnbull, *Forest People; Wayward Servants*).

3. Organization is the capacity for collective action. By this definition, lone individuals have no organization unless they act as members of a group. Groups may have greater or lesser organization, depending on such factors as whether they have full-time administrators and centralized decisionmaking. See Black, *Behavior of Law,* 85.

4. *Idem*, Chapter 5.

5. Compare Reiss ,"Control of organizational life," 303–304.

6. See, e.g., Yngvesson and Hennessey, "Small claims, complex disputes." The same pattern was found in an English court that handles claims for less than £2,000. Cain, "Where are the disputes?," 127.

A small-claims court hears only civil cases concerning relatively small stakes, such as conflicts limited to no more than $750 or $1,000 (depending on the jurisdiction). These courts were created to provide an inexpensive forum for everyday disputes between ordinary citizens.

7. Wanner, "Public ordering of private relations," Part 1, 424 and 431–432.

8. For evidence from Canada, see Hagan, "The corporate advantage."

9. Wanner, "Public ordering of private relations," Part 1, 431. For a similar pattern in England, see Cain, "Where are the disputes?," 127.

10. See, e.g., Steele, "Fraud, dispute, and the consumer"; Nader, *No Access to Law.*

11. On "lumping," see Felstiner, "Influences of social organization on dispute processing," 81; Galanter, "Why the 'haves' come out ahead," 124–125.

12. See Galanter, "Afterward"; Black, *Behavior of Law,* 86–97.

13. Wanner, "Public ordering of private relations," Part 2, 302. For the same pattern in Belgium, see Van Houtte and Langerwerf, "Administration of justice by the Fiscal Affairs Chamber of the Court of Appeal of Antwerp," 139.

14. Mayhew, *Law and Equal Opportunity,* Chapter 8.

15. Wheeler et al., "Do the 'haves' come out ahead?" Wheeler and his colleagues report a small but systematic advantage for organizations that appeal cases lost to individuals compared to individuals who appeal cases lost to organizations in both civil and criminal cases. Since they count an organization as an individual when it represents an individual, however, such as when an insurance company litigates an individual client's case, their findings undoubtedly overestimate the success of lone individuals appealing against organizations and underestimate the success of organizations appealing against lone individuals.

16. See Hagan, "The corporate advantage."

17. See, e.g., Kalven and Zeisel, *American Jury,* Chapter 5; Newman, *Conviction;* Wanner, "Public ordering of private relations," Part 2, 297. Loss of a civil case is more difficult to define than loss of a criminal case, since civil defendants may formally settle with the plaintiff without losing a judgment in court. The estimate in the text—that civil defendants lose in about half the cases—is based on Wanner's report that civil plaintiffs win a decision or record their "full satisfaction" on the court record in 50.58 percent of the cases. *Idem,* 295.

18. See, e.g., Newman, *Conviction,* 3; Mileski, "Courtroom encounters."

19. See generally Stone, *Where the Law Ends;* "Stalking the wild corporation."

In the United States, organizations are sometimes fined by regulatory agencies such as the Environmental Protection Agency or the Federal Trade Commission, and, less frequently, civil courts may order them to pay "punitive damages"—a monetary payment beyond the loss suffered by the plaintiff.

20. Maybury-Lewis, *Akwẽ-Shavante Society*, 189.

21. A noteworthy exception is the recent increase of lawsuits for compensatory damages by individuals against organizations, including those nominally against individuals (such as automobile drivers and physicans) but actually against their insurance companies. See, e.g., Lieberman, *The Litigious Society*, Chapter 2; Malott, *America's Liability Explosion*. In this respect as in others, organizations increasingly perform a function once the responsibility of families: providing compensation for misfortune of various kinds. For an elaboration, see Black, "Compensation and the social structure of misfortune," 576–581.

22. More precisely, a strong kinship system characterizes agricultural societies and declines as societies evolve to a more differentiated form. As mentioned in note 2 above, many of the earliest and simplest societies—the hunters and gatherers—appear to be more individualistic in their patterns of conflict management. See, e.g., Turnbull, *Forest People; Wayward Servants;* Lee, *!Kung San;* Woodburn, "Minimal politics."

23. See, e.g., Howell, *Manual of Nuer Law;* Peters, "Some structural aspects of the feud"; Jones, *Men of Influence in Nuristan.*

24. An exception might be the individual defendant who represented an organization when the alleged wrongdoing occurred. In such cases the organization may lend assistance—a pattern often seen in the United States when corporate executives are accused of misconduct.

It should also be recognized that all legal action against organizations themselves pertains to the conduct of individuals defined as agents of these organizations. Much individual misconduct is thereby translated into organizational misconduct, a modern form of collective liability.

25. See Hoebel, *Law of Primitive Man*, 51–55.

26. An eccentricity of the Western legal tradition is that it regards the corporate group as a kind of person, a so-called "legal person." For a brief history of this practice, see Coleman, *Asymmetric Society*, Chapter 1.

Under a regime of legal individualism, it seems, nothing but an individual is legally conceivable.

27. Again, this pattern applies specifically to simple agricultural societies.

28. See, e.g., Boulding, *Organizational Revolution;* Coleman, *Asymmetric Society.*

29. The organization of citizens in other domains, however, such as

the organization of consumers and minorities, increases the organization of legal conflicts that might otherwise have remained individualized. Class actions, which incorporate a number of plaintiffs with the same complaint into a single lawsuit, have been increasing as well. What has not occurred is an effort to organize individual litigants as such, apart from their specific concerns or complaints.

30. The following overview of the Somali system draws on Lewis, "Clanship and contract in northern Somaliland"; *Pastoral Democracy.*

The Bedouin of the Sinai Peninsula have a similar institution, known as "blood-money groups." See generally Stewart, Cases and Texts in Sinai Bedouin Law.

31. "Insult" embraces a wide range of offenses such as adultery, certain cases of physical assault, and defamation. Lewis, "Clanship and contract in northern Somaliland," 287, note 3; *Pastoral Democracy,* 162.

32. Since each member would have more to lose, it seems that smaller dia-paying groups would make a greater effort to discourage misconduct by their members. And since each member would gain more from any compensation paid to the group, smaller groups might be quicker to demand a redress of their grievances.

33. It might at first appear that legal co-ops would resemble the "protective associations" imagined by Robert Nozick in a "state of nature." See his *Anarchy, State, and Utopia,* Chapter 2. But this is not the case. Unlike Nozick's protective associations, legal co-ops would operate in a conventional legal environment and would not exercise force in the manner of a government.

34. A number of American states presently maintain so-called "victim compensation programs," by which victims of crime may receive a monetary payment from the state to help defray their losses. These payments are independent of prosecution, however, and in no sense substitute for it. In fact, frequently the payment is made when the offender's identity is unknown.

Some localities have also established "restitution programs." These require direct payments by convicted criminals to their victims. But thus far the greatest beneficiaries have been organizational victims of individual offenders. See Karmen, *Crime Victims,* 189–190.

35. See, e.g., Danzig, "Toward the creation of a complementary, decentralized system of criminal justice"; McGillis and Mullen, *Neighborhood Justice Centers;* Wahrhaftig, "Overview of community-oriented citizen dispute resolution programs in the United States."

36. See, e.g., MacCormack, "Procedures for the settlement of disputes in 'simple societies' "; Nader and Todd, *Disputing Process;* Roberts, *Order and Dispute.*

Contrary to some popular conceptions of the past, including a well-known theory of legal evolution (Durkheim, *Division of Labor*), punishment in the name of society as a whole is rare in the simplest societies.

37. Tribal justice often favors those with the strongest and most numerous connections to others willing to take their side. See, e.g., Barth, *Political Leadership among Swat Pathans*, 119–120; Gulliver, *Social Control in an African Society*, 56–58 and 116–127.

38. The anthropologist Ralph Linton refers to this as the problem of "the finally intolerable." Quoted in Llewellyn and Hoebel, *Cheyenne Way*, 49.

39. Even modern liability insurance may undermine a sense of individual responsibility in the general population. See Koch, "Liability and social structure," 125–126.

40. Moore, "Legal liability and evolutionary interpretation," 89–90 (punctuation edited). Expulsion need not be geographical, but may simply involve an abrogation of the rights and privileges of group membership.

Among the Nuristani of Afghanistan, the fellow clansmen of a wrongdoer normally pay half of the compensation to the victim's family, and the wrongdoer's family pays the other half. Compensation received from another clan is divided in the same fashion. Jones, *Men of Influence in Nuristan*, 69. But if an individual continually creates expenses for fellow clansmen, and if all efforts to pressure him into conformity fail, the clan will dissociate itself from his wrongdoing and leave his family and him to pay all the compensation themselves. As one man explained:

> If a member of our [clan] is always in trouble, if we have to pay many fines for him, then we try to make him behave. Finally, if it continues, we make him pay his own fines. He will have to give all his own goats, his own cheeses . . . everything. He will become poor (*idem*, 70).

Another traditional strategy for dealing with recidivists is assassination. See, e.g., Balikci, *Netsilik Eskimo*, 192; Westermeyer, "Assassination and conflict resolution in Laos."

41. See Moore, "Legal liability and evolutionary interpretation," 92–93.

42. Among the Shavante of Brazil, who allow only groups ("factions") to bring legal complaints, a man abandoned by his faction is virtually "outlawed"—without legal rights. Maybury-Lewis, *Akwẽ-Shavante Society*, 189.

Outlawry in this sense was also a major sanction in medieval England. Pollock and Maitland, *History of English Law,* Volume 1, 476–478.

43. This is not a new idea, however. In the 1940s, for example, the criminologist Edwin Sutherland made similar observations. See, e.g., his "White-collar criminality"; "Is white-collar crime, crime?"; and "Crime of corporations."

44. Reiss, "Control of organizational life," 302.

45. See, e.g., Stone, *Where the Law Ends;* "Stalking the wild corporation"; Nader and Shugart, "Old solutions for old problems."

46. Stone, "Stalking the wild corporation," 92–93.

47. Nader and Shugart, "Old solutions for old problems," 98.

48. Galanter, "Afterward"; Black, *Behavior of Law,* 92–97.

49. Legal co-ops might also have consequences that would be regarded as undesirable. Since organizations are more litigious than individuals, for example, legal co-ops might increase the overall rate of litigation. And since organizations are more successful than individuals when they go to court, the size of damage awards might increase. In addition, legal co-ops would seemingly reduce the participation of individuals in their own cases, a development some observers would decry. See, e.g., Christie, "Conflicts as property."

If legal co-ops come into being, their behavior will undoubtedly be closely monitored and possibly counteracted in some respects.

50. *Idem.*

51. Some readers might object that an institution suited to Somali nomads—dia-paying groups—would be difficult to transplant to a radically different setting such as modern America. But a complete transplant of dia-paying groups is not contemplated here. Though inspired by dia-paying groups, the system of legal co-operative associations outlined earlier would appear entirely compatible with contemporary life. Modern conditions generate organizations of all kinds, and people increasingly live in a state of organizational dependency. See, e.g., Stinchcombe, "Social structure and organizations," 142–155; Black, "Compensation and the social structure of misfortune," 579–581. Legal co-ops would feed on these pre-existing conditions and tendencies.

## Chapter 4

1. Selznick, "Sociology of law," 52.

2. *Idem,* 53.

3. For Selznick, "the rule of law" is synonymous with "legality" and pertains primarily to "how policies and rules are made and applied."

*Law, Society, and Industrial Justice,* 112 (italics omitted). In the handling of cases, this requires the following:

> The same rule is applied to every member of a legally defined class of cases. A particular case cannot be handled without risk to legality unless it belongs to a category, unless it can be so classified that a general rule applicable to the entire class of cases can be invoked (*idem,* 15; punctuation edited).

He further explains that legality requires "universalism" in the application of rules, meaning decisions should "transcend the special interests of persons or groups." *Idem,* 16.

4. See, e.g., Abbott, *In the Belly of the Beast,* 90.

5. This range of variation pertains only to the handling of homicides regarded as intentional and malicious. See, e.g., Lundsgaarde, *Murder in Space City,* especially 224–229.

6. See, e.g., LaFave, *Arrest,* Chapter 22; Spradley, *You Owe Yourself a Drunk;* Mileski, "Courtroom encounters." The penalty in drunkenness and prostitution cases seems to vary primarily with the criminal record of the defendant.

7. See, e.g., Barton, *Ifugao Law;* Howell, *Manual of Nuer Law;* Lewis, "Clanship and contract in northern Somaliland"; Koch, *War and Peace in Jalémó.*

The range of variation is relatively narrow in the handling of killings beyond the family but within the community. Those within the nuclear family, while not approved, may nonetheless receive no punishment or compensation, and those against members of socially distant groups may be viewed as praiseworthy. See, e.g., Howell, *idem,* 57–58 and 207–208; Middleton, *Lugbara of Uganda,* Chapter 4; Koch, *idem,* Chapters 4–6; Cooney, Social Control of Homicide, Chapter 2.

8. See Black, "Social control as a dependent variable," 17.

9. See, e.g., Wolfgang, *Patterns in Criminal Homicide;* Lundsgaarde, *Murder in Space City.*

10. See, e.g., Howell, *Manual of Nuer Law;* Lewis, "Clanship and contract"; Peters, "Some structural aspects of the feud among the camel-herding Bedouin of Cyrenaica."

11. Howell, *idem,* 57–58 and 207–208.

12. Blacks have special disadvantages associated with their own peculiar history in American society, but their major disadvantages in legal life appear to be shared with others of lower social status. See, e.g., Boris, "Stereotypes and dispositions for criminal homicide." Blacks who are wealthy and well-integrated into society (such as professionals and businesspeople) are legally better off than poor and socially marginal whites.

13. See, e.g., Bowers and Pierce, "Arbitrariness and discrimination under post-*Furman* capital statutes."

14. See, e.g., Reiss, "Measurement of the nature and amount of crime."

15. See, e.g., Chiricos and Waldo, "Socioeconomic status and criminal sentencing."

16. Burchett, Race and the AWOL Offender.

17. Racial discrimination appears to be largely absent from present-day American military life in general. See Moskos, "Success story."

18. One can obviously judge that the driver of a new and expensive sports car is likely to be socially superior to the driver of an old and dilapidated pick-up truck, but most automobiles are more homogeneous and less informative. The drivers may also differ socially from the owners. The important point, however, is that less is known about the social characteristics of parking offenders—their ethnicity, marital status, occupation, criminal record, etc.—than about the characteristics of the accused parties in face-to-face settings such as courtrooms, where direct observations, interviews, and oral presentations are possible.

19. See Black, *Manners and Customs of the Police*, 32.

20. *Idem*, 33–35.

21. Heussenstamm, "Bumper stickers and the cops."

22. See Black and Baumgartner, "On self-help in modern society," 204, note 18; Black, "Social control as a dependent variable," 19.

23. This assumes that all the other parties involved, such as lawyers and legal officials, are socially similar or otherwise neutralized as well.

24. See Black, "Social control as a dependent variable," 19–20.

25. See Black, *Behavior of Law*, 113–117.

26. See, e.g., Berman, *Justice in the U.S.S.R.*, 305–306.

27. Black, *Manners and Customs of the Police*, 6–7. This generalization excludes cases in which the caller is told that the matter will or should be referred elsewhere. *Idem*, 6, note 4.

28. See *idem*, 146.

29. See generally *idem*.

30. See Black, "Social control as a dependent variable," 19–20.

31. See, e.g., Engel, "Oven bird's song."

32. Roscoe Pound, an eminent legal scholar and former dean of Harvard Law School, regards what he calls "the socialization of law" as the highest level of legal development. But he means something entirely different from the usage in the text above. For him, law is socialized to the degree that it reflects the needs of society as a whole. See, e.g., *Jurisprudence*, Chapter 7, Section 35.

33. Aggrieved citizens also may have considerable social information about a case before they decide whether to contact a lawyer or legal official in the first place.

34. O'Barr, *Linguistic Evidence.*

35. A lesser step in the same direction would be to obscure social characteristics cosmetically, such as by requiring all courtroom participants to wear standardized robes. Other social indicators, however, such as speech, skin color, and hairstyle would remain.

36. O'Barr, *Linguistic Evidence,* Chapter 5.

37. See, e.g., Strodtbeck et al., "Social status in jury deliberations."

38. See, e.g., Newman, *Conviction;* Mileski, "Courtroom encounters."

39. See Chapter 1.

40. *Idem.*

41. Computerized justice would eliminate discretion—flexibility— from the legal process and would constitute an extreme form of what Roscoe Pound condemns as "mechanical jurisprudence" in his article so entitled.

Many legal scholars share Pound's belief that discretion is a necessary and valuable element of justice, while others believe it should be reduced as much as possible. See, e.g., Goldstein, "Police discretion not to invoke the criminal process"; Davis, *Discretionary Justice.*

42. Lawyers would probably become almost exclusively involved in advising and negotiation, functions that already dominate their daily activities.

43. From Anatole France's novel, *Le Lys Rouge,* 117–118 (author's translation).

## Chapter 5

1. E.g., Stone, *Where the Law Ends;* "Stalking the wild corporation"; Nader and Shugart, "Old solutions for old problems."

2. See generally Geis, *White-Collar Criminal.*

3. E.g., Straus et al., *Behind Closed Doors;* Dutton, "Criminal justice response to wife assault."

4. *Leviathan,* 100.

5. Recall that in this book law refers to governmental social control (see Chapter 1). Hence, stateless societies are by definition lawless as well.

6. See, e.g., Middleton and Tait, *Tribes without Rulers;* MacCormack, "Procedures for the settlement of disputes in 'simple societies' "; Roberts, *Order and Dispute.*

7. See, e.g., Diamond, *Primitive Law;* Schwartz and Miller, "Legal evolution and societal complexity"; Service, *Primitive Social Organization.*

8. See Galanter, "Reading the landscape of disputes"; Black, "Jurocracy in America."

9. This typology was developed with M. P. Baumgartner and presented jointly to Harvard Law School's Center for Criminal Justice in February 1982. For an elaboration, see Black, "Elementary forms of conflict management."

Similar typologies can be found in Koch, *War and Peace in Jalémó;* Sander, "Varieties of dispute processing"; Gulliver, *Disputes and Negotiations,* 1–3.

10. See, e.g., Rieder, "Social organization of vengeance"; Marongiu and Newman, *Vengeance.*

11. See, e.g., Black-Michaud, *Cohesive Force;* Boehm, *Blood Revenge.*

12. See, e.g., Peristiany, *Honour and Shame.*

13. See, e.g., Foucault, *Discipline and Punish.*

14. See, e.g., Baumgartner, "Social control from below."

15. E.g., Peters, "Some structural aspects of the feud among the camel-herding Bedouin of Cyrenaica"; Kiefer, *Tausug;* Koch, *War and Peace in Jalémó;* Chagnon, *Yanomamö.*

16. E.g., Bloch, *Feudal Society,* Chapter 30; Yablonsky, *Violent Gang;* Liebow, *Tally's Corner;* Abbott, *In the Belly of the Beast;* Black, "Crime as social control"; Baumgartner, *Moral Order of a Suburb,* Chapter 4.

17. E.g., Turnbull, *Wayward Servants,* 100–109; Fürer-Haimendorf, *Morals and Merit,* 17–24; Woodburn, "Minimal politics."

18. Hirschman, *Exit, Voice, and Loyalty.*

19. Morrill, Conflict Management among Corporate Executives, Chapter 2.

20. Baumgartner, "Social control in suburbia"; *Moral Order of a Suburb.*

21. See Carneiro, "Theory of the origin of the state"; Taylor, *Community, Anarchy and Liberty,* 129–139; Mann, *Sources of Social Power,* Chapters 3–4.

22. See Gulliver, *Disputes and Negotiations.*

23. E.g., Kroeber, "Law of the Yurok Indians," 514–515; Gulliver, "Dispute settlement without courts"; Engel, *Code and Custom in a Thai Provincial Court,* Chapter 5.

24. See, e.g., Newman, "Pleading guilty for consideration"; *Conviction;* Macaulay, "Lawyers and consumer protection laws"; Mather,

*Plea Bargaining or Trial?;* Griffiths, "What do Dutch lawyers actually do in divorce cases?"; Sarat and Felstiner, "Law and strategy in the divorce lawyer's office."

25. Curran, "American lawyers in the 1980s," 20.

26. For typologies of third parties, see Galtung, "Institutionalized conflict resolution"; Black and Baumgartner, "Toward a theory of the third party," 98–107.

27. See Gulliver, "On mediators"; Witty, *Mediation and Society,* Chapter 1; Merry, "Social organization of mediation in nonindustrial societies."

28. See, e.g., Gluckman, "Gossip and scandal"; Cohen, "Who stole the rabbits?"; Merry, "Rethinking gossip and scandal."

29. See, e.g., Felstiner, "Influences of social organization on dispute processing," 81; Galanter, "Why the 'haves' come out ahead," 124–125; Yngvesson, "Responses to grievance behavior"; Merry, "Going to court"; Baumgartner, "Social control in suburbia"; *Moral Order of a Suburb.*

30. See generally Black, *Behavior of Law,* Chapters 2–6.

31. See, e.g., Biderman, "Surveys of population samples for estimating crime incidence"; Garofalo and Hindelang, *Introduction to the National Crime Survey.*

32. See Black, "Common sense in the sociology of law," 20–22.

33. Trubek et al., *Civil Litigation Research Project,* S-19.

34. E.g., *idem,* S-23.

35. See, e.g., Parsons, *Social System,* Chapter 7; Lemert, "The concept of secondary deviation."

36. See generally Black, *Behavior of Law.*

37. *Idem,* 107–111. This principle is occasionally misinterpreted to mean that the overall quantity of social control is constant (e.g., Griffiths, "Division of labor in social control," 62). When law replaces another form of social control (or vice versa), however, the severity of the response to a particular grievance may change. See Black, "Social control as a dependent variable," 15, note 20. Law may entail more severity than nonlegal modes of conflict management such as avoidance or mediation, for example, whereas violent self-help may entail more severity than law. The overall quantity of social control may therefore fluctuate even while law varies inversely with other social control.

38. Pound, *Social Control through Law.*

39. See Black and Baumgartner, "On self-help in modern society," 195–196.

40. Taylor, *Community, Anarchy and Liberty,* 134.

The Hobbesian view that life without law would be "solitary, poor, nasty, brutish, and short" (*Leviathan,* 100) is arguably an intellectualization of legal dependency.

41. Gross, "Social control under totalitarianism," 67 and 69 (italics partially omitted).

42. *Idem,* 72.

43. This is not to discount the presence or importance of government-initiated social control in totalitarian societies. See, e.g., Dallin and Breslauer, *Political Terror in Communist Systems.* Moreover, it should not be inferred that nonlegal social control totally disappears in these societies. For example, family members, friends, neighbors, and workmates surely enforce some kind of moral order among themselves.

44. See, e.g., Parsons, *Societies;* Unger, *Law in Modern Society;* Luhmann, *Differentiation of Society;* Rueschemeyer, *Power and the Division of Labour.*

45. The offender temporarily returned to his car when someone in an apartment across the street called out, "Let that girl alone!" Later, in court, the man remarked that he had assumed the one who yelled at him "would close his window and go back to sleep." Brownmiller, *Against Our Will,* 217–218.

46. See Black and Baumgartner, "On self-help in modern society," 195–196.

A case similar to Kitty Genovese's recently occurred near Washington, D.C. See Churchville, "Woman raped in Greenbelt as neighbors ignore screams."

47. See Black, "Crime as social control," 39–40.

48. Le Vine, "Gusii sex offenses," 476–477.

49. Colson, *Tradition and Contract,* 40–42; see also Kiefer, "Modes of social action in armed combat," 595; Gordon and Meggitt, *Law and Order in the New Guinea Highlands,* Chapters 1–2.

50. McGrath, *Gunfighters, Highwaymen, and Vigilantes,* 161–162, 176–182, and Chapter 13. In the Western towns studied by McGrath, however, the homicide rate was extremely high. *Idem,* 253–255.

51. See, e.g., Griffiths, "What do Dutch lawyers actually do in divorce cases?"

52. See, e.g., Blankenburg et al., *Alternative Rechtsformen und Alternativen zum Recht;* Abel, *Politics of Informal Justice.* For an historical perspective on the subject, see Auerbach, *Justice without Law?*

53. See, e.g., Krimerman and Perry, *Patterns of Anarchy;* Apter and Joll, *Anarchism Today;* Baldelli, *Social Anarchism;* Bookchin, *Post-Scarcity Anarchism.*

54. See, e.g., Middleton and Tait, *Tribes without Rulers;* Roberts, *Order and Dispute.*

55. Carver, *Travels throughout North America*, 142, quoted in Miller, "Two concepts of authority," 272 (punctuation edited).

56. Lee, "Politics, sexual and non-sexual, in an egalitarian society," 49–50.

57. E.g., Tifft and Sullivan, *Struggle to Be Human;* quoted phrase on page 154.

58. Youngblood, *Expanded Cinema*, 415–419; see also Bookchin, *Post-Scarcity Anarchism.*

59. See, e.g., Zablocki, *Joyful Community*, Chapter 7; Kanter, *Commitment and Community.*

60. But see Sennett, *Uses of Disorder*, arguing that modern cities are ideal settings for anarchy. See also Rothbard, "Society without a state."

61. Black, *Behavior of Law*, Chapters 2–4.

62. See, e.g., Taylor, *Community, Anarchy and Liberty.*

63. See Black, *Behavior of Law*, Chapter 7.

64. See, e.g., Sander, "Varieties of dispute processing"; McGillis and Mullen, *Neighborhood Justice Centers;* Abel, *Politics of Informal Justice*, Volume 1; Tomasic and Feeley, *Neighborhood Justice.*

65. There is also a small movement in the United States to abolish prisons and one in Holland to abolish criminal justice itself.

66. Nozick proposes a "minimal state" as an ideal system of government. See generally *Anarchy, State, and Utopia.* Such a government would be "limited to the narrow functions of protection against force, theft, fraud, enforcement of contracts, and so on." *Idem*, ix.

67. See generally Haley, "Myth of the reluctant litigant"; "Politics of informal justice"; "Sheathing the sword of justice in Japan." Compare Kawashima, "Dispute resolution in contemporary Japan"; Miyazawa, "Taking Kawashima seriously."

68. Haley, *idem.*

69. *Idem*, "Sheathing the sword of justice in Japan," 273.

70. *Idem*, "Myth of the reluctant litigant," 381.

71. *Idem*, 369–370.

72. A policy of legal minimalism does not occur in a social vacuum. In Japan, its development has probably been encouraged by such factors as the high degree of cultural homogeneity and social cohesiveness and a robust system of informal social control. See Kawashima, "Dispute resolution in contemporary Japan."

In the United States, the new interest in delegalization may be associated partly with the breakdown of hierarchies and other kinds of inequality. See Black, "Jurocracy in America," 276–278.

73. See Haley, "Myth of the reluctant litigant," 371.

74. See Luhmann, *Sociological Theory of Law*, Chapter 2.

75. Haley, "Sheathing the sword of justice in Japan," 279.

76. *Idem.*

77. See Bayley, *Forces of Order*, Chapter 1.

Despite the higher level of dependence on law in the West, much misconduct, including illegality, is discouraged by informal systems of social control as well. See, e.g., Macaulay, "Non-contractual relations in business"; Merry, *Urban Danger*, 178–179.

Perhaps it should be added that the Japanese system of social control may include features that many Westerners would dislike. For instance, it appears to lack the means of redress available to victims of negligence and other misfortune in the West. It may also have a tendency to discourage individuality, and it may offer less protection to foreigners, strangers, and outcasts. See, e.g., Hane, *Peasants, Rebels, and Outcastes*, 139–171.

78. See, e.g., Ekland-Olson et al., "The paradoxical impact of criminal sanctions"; Reuter, "Social control in illegal markets."

In modern America, parts of the underworld also maintain their own system of third-party settlement, a service sometimes provided by the criminal syndicate known as the Mafia. Reuter, *idem*, 40–56.

79. See Black and Baumgartner, "On self-help in modern society," 195–199.

80. Bystander inaction during the commission of rape is already a criminal offense in some American localities. This occurred after a highly publicized Massachusetts case in which a woman was gang-raped on a tavern pool table while a large number of patrons watched. None interfered, and some reportedly laughed and cheered.

81. See Felstiner, "Influences of social organization on dispute processing"; Baumgartner, "Social control in suburbia"; *Moral Order of a Suburb*.

82. Modern societies such as the United States may lack a social infrastructure conducive to informal settlement by third parties. Traditional societies often have individuals available for this purpose, such as village elders, religious leaders, and family patriarchs. See, e.g., Kawashima, "Dispute resolution in contemporary Japan"; Engel, *Code and Custom in a Thai Provincial Court*, Chapter 5; Merry, "Social organization of mediation in nonindustrial societies." The absence of traditional authority may help to explain the heavy reliance on lawyers and courts in modern life.

83. For a discussion of "depolicing" as a strategy of crime control, see Black and Baumgartner, "On self-help in modern society," 195–199.

84. See, e.g., Yngvesson and Hennessey, "Small claims, complex disputes."

85. Electronic technology now facilitates better than ever the col-

lection and dissemination of information about people's credit history and past misconduct, and a regime of legal minimalism would probably lead to a more widespread application of this technology. Rothbard, "Society without a state," 199. This in turn might result in abuses of individual rights requiring their own remedies.

86. This is not to say that a Western society such as modern America could reduce its level of legal activity to Japan's. Various characteristics of the United States, such as its greater heterogeneity and atomization, would probably continue to generate a relatively high volume of legal life. See Black, *Behavior of Law,* 40–46 and 73–78.

87. See, e.g., Abel, "Delegalization," 32–42; Auerbach, *Justice without Law?,* 144–145.

88. Abel, *idem,* 41.

89. See, e.g., Werthman and Piliavin, "Gang members and the police"; Spradley, *You Owe Yourself a Drunk;* Wiseman, *Stations of the Lost.*

90. As noted in Chapter 1, capital punishment is also directed disproportionately at people who kill strangers. Gross and Mauro, "Patterns of death," 58–59. Its abolition would therefore reduce relational as well as racial and status discrimination.

91. See, e.g., Galanter, "Why the 'haves' come out ahead."

92. See, e.g., Chambliss and Seidman, *Law, Order, and Power;* Quinney, *Critique of Legal Order;* Greenberg, *Crime and Capitalism;* Beirne and Quinney, *Marxism and Law.*

93. Delegalization might effectively relocate some discrimination to informal settings, but this would not necessarily duplicate the present system of state-sponsored discrimination.

## Chapter 6

1. E.g., Maine, *Ancient Law;* Pollock and Maitland, *History of English Law;* Hall, *Theft, Law, and Society;* Horwitz, *Transformation of American Law.*

2. E.g., Schwartz and Miller, "Legal evolution and societal complexity"; Abel, "Comparative theory of dispute institutions in society"; Wimberly, "Legal evolution"; Shapiro, *Courts.*

3. E.g., Weber, *Max Weber on Law in Economy and Society;* Fallers, *Law without Precedent;* Nonet and Selznick, *Law and Society in Transition;* Sterling and Moore, "Weber's analysis of legal rationalization."

4. E.g., Durkheim, *Division of Labor in Society;* "Two laws of penal evolution"; Rusche and Kirchheimer, *Punishment and Social Structure;*

Moore, "Legal liability and evolutionary interpretation"; Spitzer, "Punishment and social organization."

5. E.g., Pound, *Social Control through Law;* Abel, "Western courts in non-Western settings"; Lieberman, *Litigious Society;* Friedman, "Courts over time."

6. Unless they are mentioned in the written law itself, such as when legal rules explicitly distinguish juveniles, married couples, or organizations from other parties.

7. See, e.g., Bowers and Pierce, "Arbitrariness and discrimination under post-*Furman* capital statutes."

8. See, e.g., Lundsgaarde, *Murder in Space City,* especially 224–229.

9. E.g., Hart, *Concept of Law;* Fuller, *Morality of Law.*

10. See Black and Baumgartner, "Toward a theory of the third party," 113; Black, "Social control as a dependent variable," 24.

11. See Toulmin, "Equity and principles."

12. See Black and Baumgartner, "Toward a theory of the third party," 113; Black, "Social control as a dependent variable," 22–23.

13. Roberts, *Order and Dispute,* 170.

14. E.g., Gillin, "Crime and punishment among the Barama River Carib of British Guiana," 334; Hoebel, *Political Organization and Law-Ways of the Comanche Indians,* 6.

15. E.g., Gulliver, "Dispute settlement without courts," 65–66; also see generally Llewellyn and Hoebel, *Cheyenne Way,* 20–29.

It has been proposed that rules are less prominent in social settings where negotiation rather than adjudication or arbitration dominates the management of conflict. Gulliver, "Case studies of law in non-Western societies," 18.

Another hypothesis is that rules are more prominent in societies where the role of judge has evolved as a specialized occupation. Fallers, *Law without Precedent,* 328–331.

16. See Stein, *Regulae Iuris;* Toulmin, "Equity and principles," 5–7.

17. This is occurring partly because organizations such as consumer- and environmental-protection groups are demanding more severity toward higher-status offenders. Often these offenders victimize the government itself, such as by overcharging or deceiving its agencies, and this too generates cases. In addition, the growth and atomization of society has depersonalized the world of social elites, and they are increasingly willing to use law—including criminal law—against each other.

18. On the social impact of complainants, see Black, "Mobilization of law"; Nader, "A user theory of law."

19. Among other factors diversifying users of law are organizations such as civil-rights and tenants' groups that sponsor and encourage legal action by poor people, the breakdown of informal hierarchies between status levels (such as once characterized race relations in the American South), and the decline of ethnic enclaves with traditional modes of conflict management that once diverted cases from the courts. See, e.g., Doo, "Dispute settlement in Chinese-American communities"; Merry, "Going to court."

20. See Lieberman, *Litigious Society*; Friedman, *Total Justice*.

21. See Galanter, "Reading the landscape of disputes"; Black, "Jurocracy in America."

22. See Cooney, "Traditions of legal sociology," 20–25.

23. See generally *idem*. The same applies to knowledge of other kinds. See, e.g., Mannheim, *Ideology and Utopia*; Schwartz, *Vertical Classification*.

24. This phrasing was suggested by Peter K. Manning.

25. For a discussion of "self-fulfilling prophesies," see Merton, *Social Theory and Social Structure*, Chapter 11.

26. Harper, *Code of Hammurabi*, Sections 203, 202, and 205, respectively.

27. See, e.g., Moore, *Power and Property in Inca Peru*; Garnsey, "Legal privilege in the Roman Empire"; Orenstein, "Toward a grammar of defilement in Hindu sacred law"; Offner, *Law and Politics in Aztec Texcoco*.

28. See, e.g., Oberg, "Crime and punishment in Tlingit society"; Howell, *Manual of Nuer Law*; Peters, "Some structural aspects of the feud among the camel-herding Bedouin of Cyrenaica"; Fallers, *Law without Precedent*; Koch, *War and Peace in Jalémó*.

29. Howell, *idem*, 207–208 and 57–58 (punctuation edited).

30. See, e.g., Pike, *History of Crime in England*, Volumes 1 and 2; Pollock and Maitland, *History of English Law*.

31. See Lieberman, *Litigious Society*, Chapter 6.

32. See Galanter, "Modernization of law."

33. Quoted phrase from Fuller, *Morality of Law*, 96.

34. The concept of sociological jurisprudence, advanced by Roscoe Pound in 1907, originally referred to the shaping of legal policies to fulfill the needs of society as a whole. See, e.g., Pound, "Survey of social interests"; *Jurisprudence*, Chapter 6; see also Geis, "Sociology and sociological jurisprudence."

For recent scholarship in this earlier tradition, see, e.g., Selznick, *Law, Society, and Industrial Justice*; Nonet and Selznick, *Law and Society in Transition*.

35. For a detailed overview of these factors, see Black, *Behavior of Law*. It should be recognized, however, that more social variables of this sort will surely be discovered in the years to come.

36. See, e.g., Gross and Mauro, "Patterns of death," 67.

37. See, e.g., Bowers and Pierce, "Arbitrariness and discrimination."

38. Randall Collins uses the phrase "age of sociology" to refer to a growing tendency of sociology to dominate the social sciences. *Conflict Sociology*, 541–549. The usage in the text, however, refers to the growing significance of sociology for society itself.

39. See, e.g., Black, *Toward a General Theory of Social Control;* "Elementary forms of conflict management."

40. See Black, "Strategy of pure sociology," 157–161; "Social control as a dependent variable," 26–29.

41. Granovetter, "Strength of weak ties"; *Getting a Job.*

42. Crane, *Invisible Colleges.*

43. Piliavin et al., "Good Samaritanism"; Ellsworth and Langer, "Staring and approach."

# References

Abbott, Jack Henry
  1981    *In the Belly of the Beast: Letters from Prison.* New York: Vintage Books, 1982.

Abel, Richard L.
  1973    "A comparative theory of dispute institutions in society." *Law and Society Review* 8: 217–347.

  1979a   "Delegalization: a critical review of its ideology, manifestations, and social consequences." Pages 27–47 in *Alternative Rechtsformen und Alternativen zum Recht. Jahrbuch für Rechtssoziologie und Rechtstheorie*, Band 6, edited by E. Blankenburg, E. Klausa, and H. Rottleuthner. Opladen: Westdentscher Verlag.

  1979b   "Western courts in non-Western settings: patterns of court use in colonial and neo-colonial Africa." Pages 167–200 in *The Imposition of Law*, edited by Sandra B. Burman and Barbara E. Harrell-Bond. New York: Academic Press.

  (editor)
  1982    *The Politics of Informal Justice.* Volume 1: *The American Experience;* Volume 2: *Comparative Studies.* New York: Academic Press.

Ames, Walter L.
  1981    *Police and Community in Japan.* Berkeley: University of California Press.

Apter, David E., and James Joll (editors)
  1971    *Anarchism Today.* Garden City: Anchor Books.

Auerbach, Jerold S.
1983    *Justice without Law?* New York: Oxford University Press.

Ayers, Edward L.
1984    *Vengeance and Justice: Crime and Justice in the 19th-Century American South.* New York: Oxford University Press.

Ayoub, Victor F.
1965    "Conflict resolution and social reorganization in a Lebanese village." *Human Organization* 24: 11–17.

Baldelli, Giovanni
1971    *Social Anarchism.* Chicago: Aldine-Atherton.

Balikci, Asen
1970    *The Netsilik Eskimo.* Garden City: Natural History Press.

Banton, Michael
1964    *The Policeman in the Community.* London: Tavistock.

Barth, Fredrik
1959    *Political Leadership among Swat Pathans.* London: Athlone Press.

Barton, Roy Franklin
1919    *Ifugao Law.* Berkeley: University of California Press, 1969.

Baumgartner, M. P.
1978    "Law and social status in colonial New Haven, 1639–1665." Pages 153–174 in *Research in Law and Sociology: An Annual Compilation of Research,* Volume 1, edited by Rita J. Simon. Greenwich: JAI Press.

1984a   "Social control from below." Pages 303–345 in *Toward a General Theory of Social Control,* Volume 1: *Fundamentals,* edited by Donald Black. Orlando: Academic Press.

1984b   "Social control in suburbia." Pages 79–103 in *Toward a General Theory of Social Control,* Volume 2: *Selected Problems,* edited by Donald Black. Orlando: Academic Press.

1985    "Law and the middle class: evidence from a suburban town." *Law and Human Behavior* 9: 3–24.

1988    *The Moral Order of a Suburb.* New York: Oxford University Press.

Bayley, David H.
1976    *Forces of Order: Police Behavior in Japan and the United States*. Berkeley: University of California Press.

Beattie, J. M.
1986    *Crime and the Courts in England, 1660–1800*. Princeton: Princeton University Press.

Beirne, Piers
1979    "Empiricism and the critique of Marxism on law and crime." *Social Problems* 26: 373–385.
1980    "Some more empiricism in the study of law: a reply to Jacobs." *Social Problems* 27: 471–475.

Beirne, Piers, and Richard Quinney (editors)
1982    *Marxism and Law*. New York: John Wiley.

Berman, Harold J.
1963    *Justice in the U.S.S.R.: An Interpretation of Soviet Law*. New York: Random House (second edition; first edition, 1950).

Berman, Jesse
1969    "The Cuban popular tribunals." *Columbia Law Review* 69: 1317–1354.

Bernard, J. L.
1979    "Interaction between the race of the defendant and that of jurors in determining verdicts." *Law and Psychology Review* 5: 103–111.

Biderman, Albert D.
1967    "Surveys of population samples for estimating crime incidence." *Annals of the American Academy of Political and Social Science* 374: 16–33.

Black, Donald
1970    "Production of crime rates." *American Sociological Review* 35: 733–748.
1971    "The social organization of arrest." *Stanford Law Review* 23: 1087–1111.
1972    "The boundaries of legal sociology." *Yale Law Journal* 81: 1086–1100.
1973    "The mobilization of law." *Journal of Legal Studies* 2: 125–149.

1976    *The Behavior of Law.* New York: Academic Press.

1979a   "Common sense in the sociology of law." *American Sociological Review* 44: 18–27.

1979b   "A note on the measurement of law." *Informationsbrief für Rechtssoziologie* 2: 92–106. Reprinted in *The Manners and Customs of the Police.* New York: Academic Press.

1979c   "A strategy of pure sociology." Pages 149–168 in *Theoretical Perspectives in Sociology,* edited by Scott G. McNall. New York: St. Martin's Press.

1980    *The Manners and Customs of the Police.* New York: Academic Press.

1983    "Crime as social control." *American Sociological Review* 48: 34–45.

1984a   "Jurocracy in America." *The Tocqueville Review—La Revue Tocqueville* 6: 273–281.

1984b   "Social control as a dependent variable." Pages 1–36 in *Toward a General Theory of Social Control,* Volume 1: *Fundamentals,* edited by Donald Black. Orlando: Academic Press.

(editor)
1984c   *Toward a General Theory of Social Control.* Volume 1: *Fundamentals;* Volume 2: *Selected Problems.* Orlando: Academic Press.

1987a   "Compensation and the social structure of misfortune." *Law and Society Review* 21: 563–584.

1987b   "The elementary forms of conflict management." Unpublished paper prepared for the Distinguished Lecturer Series, School of Justice Studies, Arizona State University, Tempe, Arizona.

1989    "Social control of the self." *Virginia Review of Sociology* 1: forthcoming.

Black, Donald, and M. P. Baumgartner
1980    "On self-help in modern society." Pages 193–208 in *The Manners and Customs of the Police,* by Donald Black. New York: Academic Press.

1983    "Toward a theory of the third party." Pages 84–114 in *Empirical Theories about Courts,* edited by Keith O. Boyum and Lynn Mather. New York: Longman.

Black, Donald, and Maureen Mileski (editors)
1973    *The Social Organization of Law.* Orlando: Academic Press.

Black-Michaud, Jacob
1975    *Cohesive Force: Feud in the Mediterranean and the Middle East.* Oxford: Basil Blackwell.

Blankenburg, E., E. Klausa, and H. Rottleuthner (editors)
1979    *Alternative Rechtsformen und Alternativen zum Recht. Jahrbuch für Rechtssoziologie und Rechtstheorie,* Band 6. Opladen: Westdentscher Verlag.

Bloch, Marc
1940    *Feudal Society.* Volume 2: *Social Classes and Political Organization.* Chicago: University of Chicago Press, 1961.

Block, Richard
1974    "Why notify the police: the victim's decision to notify the police of an assault." *Criminology* 11:555–569.

Boehm, Christopher
1984    *Blood Revenge: The Enactment and Management of Conflict in Montenegro and Other Tribal Societies.* Philadelphia: University of Pennsylvania Press, 1987.

Bohannan, Paul J.
1957    *Justice and Judgment among the Tiv.* London: Oxford University Press.

Bookchin, Murray
1971    *Post-Scarcity Anarchism.* Palo Alto: Ramparts Press.

Boris, Steven Barnet
1979    "Stereotypes and dispositions for criminal homicide." *Criminology* 17: 139–158.

Boulding, Kenneth E.
1953    *The Organizational Revolution: A Study of the Ethics of Economic Organization.* New York: Harper and Brothers.

Bowers, William J., and Glenn L. Pierce
1980    "Arbitrariness and discrimination under post-*Furman* capital statutes." *Crime and Delinquency* 26: 563–635.

Brownmiller, Susan
1975    *Against Our Will: Men, Women and Rape.* New York: Bantam Books, 1976.

Buckle, Suzann R. Thomas, and Leonard G. Buckle
1977　*Bargaining for Justice: Case Disposition and Reform in the Criminal Courts.* New York: Praeger.

Burchett, Bruce M.
1983　Race and the AWOL Offender: The Effect of the Defendant's Race on the Outcome of Courts-Martial Involving Absence without Leave. Unpublished doctoral dissertation, Department of Sociology, Carleton University, Ottawa, Canada.

Cain, Maureen, and Alan Hunt (editors)
1979　*Marx and Engels on Law.* London: Academic Press.

Cain, Maureen
1983　"Where are the disputes? A study of a first instance civil court in the U.K." Pages 119–133 in *Disputes and the Law,* edited by Maureen Cain and Kálmán Kulcsar. Budapest: Akadémiai Kiadó.

Carneiro, Robert L.
1970　"A theory of the origin of the state." *Science* 169: 733–738.

Carver, Jonathon
1797　*Travels throughout North America.* Philadelphia: Key and Simpson.

Chagnon, Napoleon A.
1977　*Yanomamö: The Fierce People.* New York: Holt, Rinehart and Winston (second edition; first edition, 1968).

Chambliss, William J., and Robert B. Seidman
1971　*Law, Order, and Power.* Reading: Addison-Wesley.

Chevigny, Paul
1969　*Police Power: Police Abuses in New York City.* New York: Vintage Books.

Chiricos, Theodore G., and Gordon P. Waldo
1975　"Socioeconomic status and criminal sentencing: an empirical assessment of a conflict proposition." *American Sociological Review* 40: 753–772.

Christie, Nils
1977　"Conflicts as property." *British Journal of Criminology* 17: 1–15.

Churchville, Victoria
  1986    "Woman raped in Greenbelt as neighbors ignore screams."
          *Washington Post,* September 19, A1 and A16.

Cohen, Eugene
  1972    "Who stole the rabbits? Crime, dispute, and social control
          in an Italian village." *Anthropological Quarterly* 45: 1–14.

Coleman, James S.
  1982    *The Asymmetric Society.* Syracuse: Syracuse University
          Press.

Collier, Jane Fishburne
  1973    *Law and Social Change in Zinacantan.* Stanford: Stanford
          University Press.

Collins, Randall
  1975    *Conflict Sociology: Toward an Explanatory Science.* New
          York: Academic Press.

Colson, Elizabeth
  1974    *Tradition and Contract: The Problem of Order.* Chicago:
          Aldine.

Cooney, Mark
  1986    "Behavioural sociology of law: a defence." *Modern Law
          Review* 49: 262–271.

  1988a   The Social Control of Homicide: A Cross-Cultural Study.
          Unpublished doctoral dissertation, Harvard Law School.

  1988b   "The traditions of legal sociology: an existential analysis."
          Unpublished paper presented at the annual meeting of the
          Law and Society Association, Vail, Colorado.

Crane, Diana
  1972    *Invisible Colleges: Diffusion of Knowledge in Scientific
          Communities.* Chicago: University of Chicago Press.

Curran, Barbara A.
  1986    "American lawyers in the 1980s: a profession in transition."
          *Law and Society Review* 20: 19–52.

Dallin, Alexander, and George W. Breslauer
  1970    *Political Terror in Communist Systems.* Stanford: Stanford
          University Press.

Dannefer, Dale
 1984    " 'Who signs the complaint?' Relational distance and the
        juvenile justice process." *Law and Society Review* 18: 249–
        271.

Danzig, Richard
 1973    "Toward the creation of a complementary, decentralized
        system of criminal justice." *Stanford Law Review* 26: 1–54.

Davis, Kenneth Culp
 1969    *Discretionary Justice: A Preliminary Inquiry.* Baton Rouge:
        Louisiana State University Press.

Diamond, A. S.
 1935    *Primitive Law.* London: Longmans, Green.

Doo, Leigh-Wai
 1973    "Dispute settlement in Chinese-American communities."
        *American Journal of Comparative Law* 21: 627–663.

Durkheim, Emile
 1893    *The Division of Labor in Society.* New York: Free Press,
        1964.

 1899–   "Two laws of penal evolution." *University of Cincinnati*
 1900    *Law Review* 38 (1969): 32–60.

Dutton, Donald G.
 1987    "The criminal justice response to wife assault." *Law and
        Human Behavior* 11: 189–206.

Dworkin, Ronald
 1977    *Taking Rights Seriously.* London: Duckworth.

Ekland-Olson, Sheldon, John Lieb, and Louis Zurcher
 1984    "The paradoxical impact of criminal sanctions: some micro-
        structural findings." *Law and Society Review* 18: 159–178.

Ellickson, Robert C.
 1986    "Of Coase and cattle: dispute resolution among neighbors
        in Shasta County." *Stanford Law Review* 38: 623–687.

Ellsworth, Phoebe C., and Ellen J. Langer
 1976    "Staring and approach: an interpretation of the stare as a
        non-specific activator." *Journal of Personality and Social
        Psychology* 33: 117–122.

Engel, David M.

1978 *Code and Custom in a Thai Provincial Court: The Interaction of Formal and Informal Systems of Justice.* Tucson: University of Arizona Press.

1984 "The oven bird's song: insiders, outsiders, and personal injuries in an American community." *Law and Society Review* 18: 551–582.

Fallers, Lloyd A.

1969 *Law without Precedent: Legal Ideas in Action in the Courts of Colonial Busoga.* Chicago: University of Chicago Press.

Feeley, Malcolm M.

1979 *The Process Is the Punishment: Handling Cases in a Lower Criminal Court.* New York: Russell Sage Foundation.

Felstiner, William L. F.

1974 "Influences of social organization on dispute processing." *Law and Society Review* 9: 63–94.

Fields, Barbara J.

1982 "Ideology and race in American history." Pages 143–177 in *Region, Race, and Reconstruction: Essays in Honor of C. Vann Woodward,* edited by J. Morgan Kousser and James M. McPherson. New York: Oxford University Press.

Fitzpatrick, Peter

1980 *Law and the State in Papua New Guinea.* London: Academic Press.

Foucault, Michel

1975 *Discipline and Punish: The Birth of the Prison.* New York: Pantheon, 1977.

France, Anatole

1923 *Le Lys Rouge.* Paris: Calmann-Levy (new edition; first edition, 1894).

Frank, Jerome

1930 *Law and the Modern Mind.* New York: Brentano's.

1949 *Courts on Trial.* Princeton: Princeton University Press.

Friedman, Lawrence M.

1983 "Courts over time: a survey of theories and research." Pages

9–50 in *Empirical Theories about Courts*, edited by Keith O. Boyum and Lynn Mather. New York: Longman.

1985    *Total Justice*. New York: Russell Sage Foundation.

Friedman, Lawrence M., and Robert V. Percival
1981    *The Roots of Justice: Crime and Punishment in Alameda County, California, 1870–1910*. Chapel Hill: University of North Carolina Press.

Fuller, Lon L.
1964    *The Morality of Law*. New Haven: Yale University Press.

Fürer-Haimendorf, Christoph von
1967    *Morals and Merit: A Study of Values and Social Controls in South Asian Societies*. Chicago: University of Chicago Press.

Galanter, Marc
1966    "The modernization of law." Pages 153–165 in *Modernization: The Dynamics of Growth*, edited by Myron Weiner. New York: Basic Books.

1974    "Why the 'haves' come out ahead: speculations on the limits of legal change." *Law and Society Review* 9: 95–160.

1975    "Afterward: explaining litigation." *Law and Society Review* 9: 346–368.

1983    "Reading the landscape of disputes: what we know and don't know (and think we know) about our allegedly contentious and litigious society." *UCLA Law Review* 31: 4–71.

Galtung, Johan
1965    "Institutionalized conflict resolution: a theoretical paradigm." *Journal of Peace Research* 2: 349–397.

Garfinkel, Harold
1949    "Research note on inter- and intra-racial homicides." *Social Forces* 27: 369–381.

Garnsey, Peter
1968    "Legal privilege in the Roman Empire." *Past and Present* 41: 3–24.

Garofalo, James, and Michael J. Hindelang
1977    *An Introduction to the National Crime Survey*. Washington, D.C.: U.S. Department of Justice, Law Enforcement Assis-

tance Administration, National Criminal Justice Information and Statistics Service.

Geis, Gilbert
1964 "Sociology and sociological jurisprudence: admixture of lore and law." *Kentucky Law Journal* 52: 267–293.
(editor)
1968 *White-Collar Criminal: The Offender in Business and the Professions.* New York: Atherton Press.

Gibbs, James L., Jr.
1963 "The Kpelle moot: a therapeutic model for the informal settlement of disputes." *Africa* 33: 1–10.

Gillin, John
1934 "Crime and punishment among the Barama River Carib of British Guiana." *American Anthropologist* 36: 331–344.

Gluckman, Max
1955 *The Judicial Process among the Barotse of Northern Rhodesia.* Manchester: Manchester University Press.
1963 "Gossip and scandal." *Current Anthropology* 4: 307–316.

Goldstein, Joseph
1960 "Police discretion not to invoke the criminal process: low-visibility decisions in the administration of justice." *Yale Law Journal* 69: 543–594.

Gordon, Robert J., and Mervyn J. Meggitt
1985 *Law and Order in the New Guinea Highlands: Encounters with Enga.* Hanover: University Press of New England.

Gottfredson, Michael R., and Michael J. Hindelang
1979 "A study of *The Behavior of Law.*" *American Sociological Review* 44: 3–18.

Gough, E. Kathleen
1960 "Caste in a Tanjore village." Pages 11–60 in *Aspects of Caste in South India, Ceylon and North-West Pakistan,* edited by E. R. Leach. Cambridge: Cambridge University Press.

Granovetter, Mark S.
1973 "The strength of weak ties." *American Journal of Sociology* 78: 1360–1380.

1974     *Getting a Job: A Study of Contacts and Careers.* Cambridge: Harvard University Press.

Greenberg, David F.
  (editor)
  1981     *Crime and Capitalism: Readings in Marxist Criminology.* Palo Alto: Mayfield.

  1982–    "Donald Black's sociology of law: a critique." *Law and*
  1983     *Society Review* 17: 337–368.

Griffiths, John
  1984     "The division of labor in social control." Pages 37–70 in *Toward a General Theory of Social Control,* Volume 1: *Fundamentals,* edited by Donald Black. Orlando: Academic Press.

  1986     "What do Dutch lawyers actually do in divorce cases?" *Law and Society Review* 20: 135–175.

Gross, Jan T.
  1984     "Social control under totalitarianism." Pages 59–77 in *Toward a General Theory of Social Control,* Volume 2: *Selected Problems,* edited by Donald Black. Orlando: Academic Press.

Gross, Samuel R., and Robert Mauro
  1984     "Patterns of death: an analysis of racial disparities in capital sentencing and homicide victimization." *Stanford Law Review* 37: 27–153.

Gulliver, Philip H.
  1963     *Social Control in an African Society: A Study of the Arusha, Agricultural Masai of Northern Tanganyika.* Boston: Boston University Press.

  1969a    "Case studies of law in non-Western societies: introduction." Pages 11–23 in *Law in Culture and Society,* edited by Laura Nader. Chicago: Aldine.

  1969b    "Dispute settlement without courts: the Ndendeuli of southern Tanzania." Pages 24–68 in *Law in Culture and Society,* edited by Laura Nader. Chicago: Aldine.

  1977     "On mediators." Pages 15–52 in *Social Anthropology and Law,* edited by Ian Hamnett. London: Academic Press.

  1979     *Disputes and Negotiations: A Cross-Cultural Perspective.* New York: Academic Press.

Hagan, John
  1982    "The corporate advantage: a study of the involvement of
          corporate and individual victims in a criminal justice sys-
          tem." *Social Forces* 60: 993–1022.

Haley, John O.
  1978    "The myth of the reluctant litigant." *Journal of Japanese
          Studies* 4: 359–390.

  1982a   "The politics of informal justice: the Japanese experience,
          1922–1942." Pages 125–147 in *The Politics of Informal
          Justice*, Volume 2: *Comparative Studies*, edited by Richard
          L. Abel. New York: Academic Press.

  1982b   "Sheathing the sword of justice in Japan: an essay on law
          without sanctions." *Journal of Japanese Studies* 8: 265–281.

Hall, Jerome
  1952    *Theft, Law, and Society.* Indianapolis: Bobbs-Merrill (sec-
          ond edition; first edition, 1935).

Hanawalt, Barbara A.
  1979    *Crime and Conflict in English Communities, 1300–1348.*
          Cambridge: Harvard University Press.

Hane, Mikiso
  1982    *Peasants, Rebels, and Outcastes: The Underside of Modern
          Japan.* New York: Pantheon Books.

Harper, Robert Francis (translator)
  1904    *The Code of Hammurabi, King of Babylon: About 2250
          B.C.* Chicago: University of Chicago Press.

Hart, H. L. A.
  1961    *The Concept of Law.* Oxford: Clarendon Press.

Hasluck, Margaret
  1954    *The Unwritten Law in Albania.* Cambridge: Cambridge
          University Press.

Hastie, Reid, Steven D. Penrod, and Nancy Pennington
  1983    *Inside the Jury.* Cambridge: Harvard University Press.

Hay, Douglas, Peter Linebaugh, John G. Rule, E. P. Thompson, and
Cal Winslow
  1975    *Albion's Fatal Tree: Crime and Society in Eighteenth-
          Century England.* New York: Pantheon.

Heinz, John T., and Edward O. Laumann
1982 *Chicago Lawyers: The Social Structure of the Bar.* New York: Russell Sage Foundation.

Hembroff, Larry A.
1987 "The seriousness of acts and social contexts: a test of Black's theory of the behavior of law." *American Journal of Sociology* 93: 322–347.

Heussenstamm, F. K.
1971 "Bumper stickers and the cops." *Trans-Action* 8: 32–33.

Hindus, Michael S.
1980 *Prison and Plantation: Crime, Justice and Authority in Massachusetts and South Carolina, 1767–1878.* Chapel Hill: University of North Carolina Press.

Hirschman, Albert O.
1970 *Exit, Voice, and Loyalty: Responses to Decline in Firms, Organizations, and States.* Cambridge: Harvard University Press.

Hobbes, Thomas
1651 *Leviathan: Or the Matter, Forme and Power of a Commonwealth Ecclesiasticall and Civil.* New York: Macmillan, 1962.

Hoebel, E. Adamson
1940 *The Political Organization and Law-Ways of the Comanche Indians.* Memoirs of the American Anthropological Association, Number 54. Menasha: American Anthropological Association.

1954 *The Law of Primitive Man: A Study in Comparative Legal Dynamics.* Cambridge: Harvard University Press.

Holmes, Oliver Wendell
1897 "The path of the law." *Harvard Law Review* 10: 457–478.

Holmstrom, Lynda Lytle, and Ann Wolbert Burgess
1978 *The Victim of Rape: Institutional Reactions.* New Brunswick: Tranaction Books, 1983.

Horwitz, Allan V.
1982– "Resistance to innovation in the sociology of law: a reply
1983 to Greenberg." *Law and Society Review* 17: 369–384.

Horwitz, Morton J.
1977    *The Transformation of American Law, 1780–1860.* Cambridge: Harvard University Press.

Howell, P. P.
1954    *A Manual of Nuer Law: Being an Account of Customary Law, Its Evolution and Development in the Courts Established by the Sudan Government.* London: Oxford University Press.

Hunt, Alan
1983    "Behavioural sociology of law: a critique of Donald Black." *Journal of Law and Society* 10: 19–46.

Inglis, Amirah
1974    *'Not a White Woman Safe': Sexual Anxiety and Politics in Port Moresby, 1920–1934.* Canberra: Australian National University Press.

Johnson, G. B.
1941    "The Negro and crime." *Annals of the American Academy of Political and Social Science* 271: 93–104.

Jones, Schuyler
1974    *Men of Influence in Nuristan: A Study of Social Control and Dispute Settlement in Waigal Valley, Afghanistan.* New York: Seminar Press.

Kagan, Richard L.
1981    *Lawsuits and Litigants in Castile, 1500–1700.* Chapel Hill: University of North Carolina Press.

Kairys, David (editor)
1982    *The Politics of Law: A Progressive Critique.* New York: Pantheon.

Kalogeropoulos, Dimitri, and Danielle Rivière
1983    "Police station discourse." Pages 69–83 in *Disputes and the Law,* edited by Maureen Cain and Kálmán Kulcsár. Budapest: Académiai Kiadó.

Kalven, Harry, Jr., and Hans Zeisel
1966    *The American Jury.* Boston: Little, Brown.

Kanter, Rosabeth Moss
1972    *Commitment and Community: Communes and Utopias in Sociological Perspective.* Cambridge: Harvard University Press.

Karmen, Andrew
  1984  *Crime Victims: An Introduction to Victimology.* Monterey: Brooks/Cole.

Kawashima, Takeyoshi
  1963  "Dispute resolution in contemporary Japan." Pages 41–72 in *Law in Japan: The Legal Order in a Changing Society,* edited by Arthur T. von Mehren. Cambridge: Harvard University Press.

Kennedy, Duncan
  1983  *Legal Education and the Reproduction of Hierarchy: A Polemic against the System.* Cambridge: Afar.

Kiefer, Thomas M.
  1970  "Modes of social action in armed combat: affect, tradition and reason in Tausug private warfare." *Man* 5: 586–596.

  1972  *The Tausug: Violence and Law in a Philippine Moslem Society.* New York: Holt, Rinehart and Winston.

Koch, Klaus-Friedrich
  1974  *War and Peace in Jalémó: The Management of Conflict in Highland New Guinea.* Cambridge: Harvard University Press.

  1984  "Liability and social structure." Pages 95–129 in *Toward a General Theory of Social Control,* Volume 1: *Fundamentals,* edited by Donald Black. Orlando: Academic Press.

Krimerman, Leonard I., and Lewis Perry (editors)
  1966  *Patterns of Anarchy: A Collection of Writings in the Anarchist Tradition.* Garden City: Anchor Books.

Kroeber, A. L.
  1926  "Law of the Yurok Indians." Pages 511–516 in *Proceedings of the 22nd International Congress of Americanists,* Volume 2. Rome: Instituto Christoforo Colombo.

Kruttschnitt, Candace
  1980–  "Social status and the sentences of female offenders." *Law*
  1981   *and Society Review* 15: 247–265.

  1982  "Women, crime and dependency: an application of the theory of law." *Criminology* 19: 495–513.

Ladinsky, Jack
  1963  "Careers of lawyers, law practice, and legal institutions." *American Sociological Review* 28: 47–54.

LaFave, Wayne R.
1965    *Arrest: The Decision to Take a Suspect into Custody.* Boston: Little, Brown.

Landy, David, and Elliot Aronson
1969    "The influence of the character of the criminal and his victim on the decisions of simulated jurors." *Journal of Experimental Social Psychology* 5: 141–152.

Lee, Richard
1979    *The !Kung San: Men, Women, and Work in a Foraging Society.* Cambridge: Cambridge University Press.

1982    "Politics, sexual and non-sexual, in an egalitarian society." Pages 37–59 in *Politics and History in Band Societies,* edited by Eleanor Leacock and Richard Lee. Cambridge: Cambridge University Press.

Lefstein, Norman, Vaughan Stapleton, and Lee Teitelbaum
1969    "In search of juvenile justice: *Gault* and its implementation." *Law and Society Review* 3: 491–562.

Lemert, Edwin M.
1967    "The concept of secondary deviation." Pages 40–64 in *Human Deviance, Social Problems, and Social Control.* Englewood Cliffs: Prentice-Hall.

Levin, Martin A.
1974    "Urban politics and judicial behavior." *Journal of Legal Studies* 3: 339–375.

Le Vine, Robert A.
1959    "Gusii sex offenses: a study in social control." *American Anthropologist* 61: 965–990.

Lewis, I. M.
1959    "Clanship and contract in northern Somaliland." *Africa* 29: 274–293.

1961    *A Pastoral Democracy: A Study of Pastoralism and Politics among the Northern Somali of the Horn of Africa.* London: Oxford University Press.

Lieberman, Jethro K.
1981    *The Litigious Society.* New York: Basic Books.

Liebow, Elliot
1967    *Tally's Corner: A Study of Negro Streetcorner Men.* Boston: Little, Brown.

Lintott, A. W.
  1968  *Violence in Republican Rome*. London: Oxford University Press.

Livingston, Debra
  1982  " 'Round and 'round the bramble bush: from legal realism to critical legal scholarship." *Harvard Law Review* 95: 1669–1690.

Llewellyn, Karl N.
  1962  *Jurisprudence: Realism in Theory and Practice*. Chicago: University of Chicago Press.

Llewellyn, Karl N., and E. Adamson Hoebel
  1941  *The Cheyenne Way: Conflict and Case Law in Primitive Jurisprudence*. Norman: University of Oklahoma Press.

Luhmann, Niklas
  1972  *A Sociological Theory of Law*. London: Routledge and Kegan Paul, 1985.

  1982  *The Differentiation of Society*. New York: Columbia University Press.

Lundsgaarde, Henry P.
  1977  *Murder in Space City: A Cultural Analysis of Houston Homicide Patterns*. New York: Oxford University Press.

Macaulay, Stewart
  1963  "Non-contractual relations in business: a preliminary study." *American Sociological Review* 28: 55–67.

  1979  "Lawyers and consumer protection laws." *Law and Society Review* 14: 115–171.

MacCormack, Geoffrey
  1976  "Procedures for the settlement of disputes in 'simple societies.' " *The Irish Jurist* 11 (new series): 175–188.

Maine, Henry Sumner
  1861  *Ancient Law: Its Connection with the Early History of Society and Its Relation to Modern Ideas*. Boston: Beacon Press, 1963.

Malott, Robert H.
  1985  *America's Liability Explosion: Can We Afford the Cost?* Chicago: FMC Corporation.

Mann, Michael
1986    *The Sources of Social Power.* Volume 1: *A History of Power from the Beginning to A.D. 1760.* Cambridge: Cambridge University Press.

Mannheim, Karl
1936    *Ideology and Utopia: An Introduction to the Sociology of Knowledge.* New York: Harcourt, Brace and World (expanded edition; original edition, 1920).

Manning, Peter K.
1977    *Police Work: The Social Organization of Policing.* Cambridge: MIT Press.

Marongiu, Pietro, and Graeme Newman
1987    *Vengeance: The Fight against Injustice.* Totowa: Rowman and Littlefield.

Mather, Lynn
1979    *Plea Bargaining or Trial? The Process of Criminal Case Disposition.* Lexington: Lexington Books.

Maybury-Lewis, David
1967    *Akwẽ-Shavante Society.* Oxford: Clarendon Press.

Mayhew, Leon
1968    *Law and Equal Opportunity: A Study of the Massachusetts Commission against Discrimination.* Cambridge: Harvard University Press.

Mayhew, Leon, and Albert J. Reiss, Jr.
1969    "The social organization of legal contacts." *American Sociological Review* 34: 309–318.

Maynard, Douglas W.
1984    *Inside Plea Bargaining: The Language of Negotiation.* New York: Plenum Press.

McGillis, Daniel, and Joan Mullen
1977    *Neighborhood Justice Centers: An Analysis of Potential Models.* Washington, D.C.: U.S. Government Printing Office.

McGrath, Roger D.
1984    *Gunfighters, Highwaymen, and Vigilantes: Violence on the Frontier.* Berkeley: University of California Press.

McIntyre, Jennie
  1967    "Public attitudes toward crime and law enforcement." *Annals of the American Academy of Political and Social Science* 374: 34–46.

Merry, Sally Engle
  1979    "Going to court: strategies of dispute management in an American urban neighborhood." *Law and Society Review* 13: 891–925.

  1981    *Urban Danger: Life in a Neighborhood of Strangers.* Philadelphia: Temple University Press.

  1982    "The social organization of mediation in nonindustrial societies: implications for informal community justice in America." Pages 17–45 in *The Politics of Informal Justice,* Volume 2: *Comparative Studies,* edited by Richard L. Abel. New York: Academic Press.

  1984    "Rethinking gossip and scandal." Pages 271–302 in *Toward a General Theory of Social Control,* Volume 1: *Fundamentals,* edited by Donald Black. Orlando: Academic Press.

Merton, Robert K.
  1957    *Social Theory and Social Structure.* Glencoe: The Free Press (revised and enlarged edition; first edition, 1949).

Middleton, John
  1965    *The Lugbara of Uganda.* New York: Holt, Rinehart and Winston.

Middleton, John, and David Tait (editors)
  1958    *Tribes without Rulers: Studies in African Segmentary Systems.* New York: Humanities Press, 1970.

Mileski, Maureen
  1971a    "Courtroom encounters: an observation study of a lower criminal court." *Law and Society Review* 5: 473–538.

  1971b    Policing Slum Landlords: An Observation Study of Administrative Control. Unpublished doctoral dissertation, Department of Sociology, Yale University.

Miller, Dorothy, and Michael Schwartz
  1966    "County lunacy commission hearings: some observations of commitments to a state mental hospital." *Social Problems* 14: 26–35.

Miller, Walter B.
  1955   "Two concepts of authority." *American Anthropologist* 57: 271–289.

Miyazawa, Setsuo
  1987   "Taking Kawashima seriously: a review of Japanese research on Japanese legal consciousness and disputing behavior." *Law and Society Review* 21: 219–241.

Moore, Sally Falk
  1958   *Power and Property in Inca Peru.* New York: Columbia University Press.

  1972   "Legal liability and evolutionary interpretation: some aspects of strict liability, self-help and collective responsibility." Pages 51–107 in *The Allocation of Responsibility*, edited by Max Gluckman. Manchester: Manchester University Press.

Morrill, Calvin Keith
  1986   Conflict Management among Corporate Executives: A Comparative Study. Unpublished doctoral dissertation, Department of Sociology, Harvard University.

Moskos, Charles C.
  1986   "Success story: blacks in the Army." *The Atlantic* 257: 64–72.

Myers, Martha A.
  1980   "Predicting the behavior of law: a test of two models." *Law and Society Review* 14: 835–857.

Myrdal, Gunnar (with the assistance of Richard Sterner and Arnold Rose)
  1944   *An American Dilemma: The Negro Problem and Modern Democracy.* New York: Harper and Brothers.

Nader, Laura
  1964   "An analysis of Zapotec law cases." *Ethnology* 3: 404–419.
  (editor)
  1980   *No Access to Law: Alternatives to the American Judicial System.* New York: Academic Press.

  1984   "A user theory of law." *Southwestern Law Journal* 38: 951–963.

Nader, Laura, and Christopher Shugart
  1980   "Old solutions for old problems." Pages 57–110 in *No Ac-*

*cess to Law: Alternatives to the American Judicial System,*
edited by Laura Nader. New York: Academic Press.

Nader, Laura, and Harry F. Todd, Jr. (editors)
1978    *The Disputing Process—Law in Ten Societies.* New York:
Columbia University Press.

Nagel, Stuart S.
1962    "Judicial backgrounds and criminal cases." *Journal of Criminal Law, Criminology and Police Science* 53: 333–339.

Nelson, William E.
1981    *Dispute and Conflict Resolution in Plymouth County, Massachusetts, 1725–1825.* Chapel Hill: University of North
Carolina Press.

Newman, Donald J.
1956    "Pleading guilty for consideration: a study of bargain justice." *Journal of Criminal Law, Criminology and Police
Science* 46: 780–790.

1966    *Conviction: The Determination of Guilt or Innocence without Trial.* Boston: Little, Brown.

Nonet, Philippe, and Philip Selznick
1978    *Law and Society in Transition: Toward Responsive Law.*
New York: Harper and Row.

Nozick, Robert
1974    *Anarchy, State, and Utopia.* New York: Basic Books.

O'Barr, William M.
1982    *Linguistic Evidence: Language, Power, and Strategy in the
Courtroom.* New York: Academic Press.

Oberg, Kalervo
1934    "Crime and punishment in Tlingit society." *American Anthropologist* 36: 145–156.

Offner, Jerome A.
1983    *Law and Politics in Aztec Texcoco.* Cambridge: Cambridge
University Press.

Orenstein, Henry
1968    "Toward a grammar of defilement in Hindu sacred law."
Pages 115–131 in *Structure and Change in Indian Society,*
edited by Milton Singer and Bernard S. Cohn. New York:
Wenner-Gren Foundation.

Parsons, Talcott
  1951    *The Social System*. New York: The Free Press.
  1966    *Societies: Evolutionary and Comparative Perspectives*. Englewood Cliffs: Prentice-Hall.

Peristiany, J. G. (editor)
  1966    *Honour and Shame: The Values of Mediterranean Society*. Chicago: University of Chicago Press.

Peters, E. L.
  1967    "Some structural aspects of the feud among the camel-herding Bedouin of Cyrenaica." *Africa* 37: 261–282.

Pike, Luke Owen
  1873    *A History of Crime in England, Illustrating the Changes of the Laws in the Progress of Civilization*, Volume 2: *From the Roman Invasion to the Accession of Henry VII*. London: Smith, Elder.
  1876    *A History of Crime in England, Illustrating the Changes of the Laws in the Progress of Civilization*, Volume 2. *From the Accession of Henry VII to the Present Time*. London: Smith, Elder.

Piliavin, Irving M., Judith Rodin, and Jane Allyn Piliavin
  1969    "Good Samaritanism: an underground phenomenon?" *Journal of Personality and Social Psychology* 13: 289–299.

Pollock, Frederick, and Frederic William Maitland
  1898    *The History of English Law: Before the Time of Edward I*. Two Volumes. Cambridge: Cambridge University Press (second edition; first edition, 1895).

Pound, Roscoe
  1908    "Mechanical jurisprudence." *Columbia Law Review* 8: 605–623.
  1910    "Law in books and law in action." *American Law Review* 44: 12–34.
  1942    *Social Control through Law*. New Haven: Yale University Press.
  1943    "A survey of social interests." *Harvard Law Review* 57: 1–39.
  1959    *Jurisprudence*. Volume 1. St. Paul: West Publishing Company.

Punch, Maurice
   1979    Policing the Inner City: A Study of Amsterdam's Warmoes-
           straat. Hamden: Archon Books.

Quinney, Richard
   1974    Critique of Legal Order: Crime Control in Capitalist So-
           ciety. Boston: Little, Brown.

Reid, John Phillip
   1970    A Law of Blood: The Primitive Law of the Cherokee Na-
           tion. New York: New York University Press.

Reiss, Albert J., Jr.
   1967    "Measurement of the nature and amount of crime." Pages
           1–183 in Studies in Crime and Law Enforcement in Major
           metropolitan Areas. A Report to the President's Commission
           on Law Enforcement and Administration of Justice. Volume
           1. Washington, D.C.: U.S. Government Printing Office.

   1971    The Police and the Public. New Haven: Yale University
           Press.

   1985    "The control of organizational life." Pages 294–308 in Per-
           spectives in Criminal Law: Essays in Honour of John Ll. J.
           Edwards, edited by Anthony N. Doob and Edward L.
           Greenspan. Aurora: Canada Law Book.

Reuter, Peter
   1984    "Social control in illegal markets." Pages 29–58 in Toward
           a General Theory of Social Control, Volume 2: Selected
           Problems, edited by Donald Black. Orlando: Academic
           Press.

Rieder, Jonathan
   1984    "The social organization of vengeance." Pages 131–162 in
           Toward a General Theory of Social Control, Volume 1:
           Fundamentals, edited by Donald Black. Orlando: Aca-
           demic Press.

Roberts, Simon
   1979    Order and Dispute: An Introduction to Legal Anthropology.
           New York: Penguin Books.

Roberts, W. Lewis
   1922    "The unwritten law." Kentucky Law Journal 10: 45–52.

Rodell, Fred
   1939    Woe unto You, Lawyers! New York: Berkeley, 1980.

Rosenthal, Douglas E.
  1974    *Lawyer and Client: Who's in Charge?* New York: Russell
          Sage Foundation.

Ross, H. Laurence
  1970    *Settled Out of Court: The Social Process of Insurance Claims
          Adjustments.* Chicago: Aldine.

Rothbard, Murray N.
  1978    "Society without a state." Pages 191–207 in *Anarchism,*
          edited by J. Roland Pennock and John W. Chapman.
          *Nomos,* Volume 19. New York: New York University Press.

Rothenberger, John E.
  1978    "The social dynamics of dispute settlement in a Sunni Mus-
          lim village in Lebanon." Pages 152–180 in *The Disputing
          Process—Law in Ten Societies,* edited by Laura Nader and
          Harry F. Todd, Jr. New York: Columbia University Press.

Rubin, Sol
  1966    "Disparity and equality of sentences—a constitutional chal-
          lenge." *Federal Rules Decisions* 40: 55–78.

Rueschemeyer, Dietrich
  1986    *Power and the Division of Labour.* Stanford: Stanford Uni-
          versity Press.

Ruggiero, Guido
  1980    *Violence in Early Renaissance Venice.* New Brunswick:
          Rutgers University Press.

Rusche, Georg, and Otto Kirchheimer
  1939    *Punishment and Social Structure.* New York: Russell and
          Russell, 1968.

Samaha, Joel
  1974    *Law and Order in Historical Perspective: The Case of Eliza-
          bethan Essex.* New York: Academic Press.

Sander, Frank E. A.
  1976    "Varieties of dispute processing." *Federal Rules Decisions*
          70: 111–133.

Sarat, Austin, and William L. F. Felstiner
  1986    "Law and strategy in the divorce lawyer's office." *Law and
          Society Review* 20: 93–134.

Schwartz, Barry
   1981    *Vertical Classification: A Study in Structuralism and the So-
           ciology of Knowledge.* Chicago: University of Chicago Press.

Schwartz, Richard D., and James C. Miller
   1964    "Legal evolution and societal complexity." *American Jour-
           nal of Sociology* 70: 159–169.

Selznick, Philip
   1968    "The sociology of law." Pages 50–59 in *International En-
           cyclopedia of the Social Sciences*, Volume 9, edited by
           David L. Sills. New York: The Free Press.

Selznick, Philip (with the assistance of Philippe Nonet and Howard
M. Vollmer)
   1969    *Law, Society, and Industrial Justice.* New York: Russell
           Sage Foundation.

Sennett, Richard
   1970    *The Uses of Disorder: Personal Identity and City Life.* New
           York: Random House, 1971.

Service, Elman R.
   1971    *Primitive Social Organization: An Evolutionary Perspective.*
           New York: Random House (second edition; first edition,
           1962).

Shapiro, Martin
   1980    *Courts: A Comparative and Political Analysis.* Chicago:
           University of Chicago Press.

Silberman, Matthew
   1985    *The Civil Justice Process: A Detroit Area Study.* Orlando:
           Academic Press.

Simmel, Georg
   1908    *The Sociology of Georg Simmel*, edited by Kurt H. Wolff.
           New York: The Free Press, 1960.

Skolnick, Jerome H.
   1966    *Justice without Trial: Law Enforcement in Democratic So-
           ciety.* New York: John Wiley.

   1967    "Social control in the adversary system." *Journal of Conflict
           Resolution* 11: 52–70.

Smith, Douglas A., and Jody R. Klein
   1984    "Police control of interpersonal disputes." *Social Problems*
           31: 468–481.

Snyder, Francis G.
   1981    *Capitalism and Legal Change: An African Transformation.*
           New York: Academic Press.

Spitzer, Steven
   1975    "Punishment and social organization: a study of Durkheim's
           theory of penal evolution." *Law and Society Review* 9:
           613–635.

Spradley, James P.
   1970    *You Owe Yourself a Drunk: An Ethnography of Urban No-
           mads.* Boston: Little, Brown.

*Stanford Law Review* Editors
   1984    "Critical Legal Studies Symposium." *Stanford Law Review*
           36 (January).

Staples, William G.
   1984    "Toward a structural perspective on gender bias in the
           juvenile court." *Sociological Perspectives* 27: 349–367.

Starr, June
   1978a   *Dispute and Settlement in Rural Turkey: An Ethnography
           of Law.* Leiden: E. J. Brill.

   1978b   "Turkish village disputing behavior." Pages 122–151 in
           *The Disputing Process—Law in Ten Societies,* edited by
           Laura Nader and Harry F. Todd, Jr. New York: Columbia
           University Press.

Steele, Eric H.
   1975    "Fraud, dispute, and the consumer: responding to con-
           sumer complaints." *University of Pennsylvania Law Review*
           123: 1107–1186.

Stein, Peter
   1966    *Regulae Iuris: From Juristic Rules to Legal Maxims.* Edin-
           burgh: Edinburgh University Press.

Sterling, Joyce S., and Wilbert E. Moore
   1987    "Weber's analysis of legal rationalization: a critique and
           constructive modification." *Sociological Forum* 2: 67–89.

Stewart, Frank H.
    1982    Cases and Texts in Sinai Bedouin Law. Unpublished manuscript.

Stinchcombe, Arthur L.
    1965    "Social structure and organizations." Pages 142–193 in *Handbook of Organizations*, edited by James G. March. Chicago: Rand McNally.

Stone, Christopher
    1975    *Where the Law Ends: The Social Control of Corporate Behavior.* New York: Harper and Row.

    1976    "Stalking the wild corporation." *Working Papers for a New Society* 4: 17ff.

Straus, Murray A., Richard T. Gelles, and Suzanne K. Steinmetz
    1980    *Behind Closed Doors: Violence in the American Family.* Garden City: Anchor Press.

Strodtbeck, Fred L., Rita M. James, and Charles Hawkins
    1957    "Social status in jury deliberations." *American Sociological Review* 22: 713–719.

Sutherland, Edwin H.
    1940    "White-collar criminality." *American Sociological Review* 5: 1–12.

    1945    "Is white-collar crime, crime?" *American Sociological Review* 10: 132–139.

    1948    "Crime of corporations." Pages 78–96 in *The Sutherland Papers*, edited by Albert K. Cohen, Alfred Lindesmith, and Karl Schuessler. Bloomington: Indiana University Press.

Tanner, R. E. S.
    1966    "The selective use of legal systems in East Africa." *East African Institute of Social Research Conference Papers:* Part E, Number 393.

Taylor, Michael
    1982    *Community, Anarchy and Liberty.* Cambridge: Cambridge University Press.

Thomas-Buckle, Suzann R., and Leonard G. Buckle
    1982    "Doing unto others: disputes and dispute processing in an urban American neighborhood." Pages 78–90 in *Neighbor-*

*hood Justice: Assessment of an Emerging Idea*, edited by Roman Tomasic and Malcolm M. Feeley. New York: Longman.

Tifft, Larry, and Dennis Sullivan
1980    *The Struggle to Be Human: Crime, Criminology, and Anarchism.* Orkney: Cienfuegos Press.

Todd, Harry F., Jr.
1978    "Litigious marginals: character and disputing in a Bavarian village." Pages 86–121 in *The Disputing Process—Law in Ten Societies*, edited by Laura Nader and Harry F. Todd, Jr. New York: Columbia University Press.

Tomasic, Roman
1985    *The Sociology of Law.* Beverly Hills: Sage.

Tomasic, Roman, and Malcolm M. Feeley (editors)
1982    *Neighborhood Justice: Assessment of an Emerging Idea.* New York: Longman.

Toulmin, Stephen
1982    "Equity and principles." *Osgoode Hall Law Journal* 20: 1–17.

Trubek, David M., Joel B. Grossman, William L. F. Felstiner, Herbert M. Kritzer, and Austin Sarat
1983    *Civil Litigation Research Project: Final Report.* Part A. Madison: University of Wisconsin Law School.

Turnbull, Colin M.
1961    *The Forest People.* New York: Simon and Schuster.

1965    *Wayward Servants: The Two Worlds of the African Pygmies.* Garden City: Natural History Press.

Twining, William
1973    *Karl Llewellyn and the Realist Movement.* London: Weidenfeld and Nicolson.

Uhlman, Thomas M.
1979    *Racial Justice: Black Judges and Defendants in an Urban Trial Court.* Lexington: Lexington Books.

Unger, Roberto Mangabeira
1976    *Law in Modern Society: Toward a Criticism of Social Theory.* New York: The Free Press.

Vago, Steven
    1981    *Law and Society.* Englewood Cliffs: Prentice-Hall.

van der Sprenkel, Sybille
    1962    *Legal Institutions in Manchu China: A Sociological Analy-
            sis.* New York: Humanities Press, 1966.

Van Houtte, Jean, and Étienne Langerwerf
    1983    "The administration of justice by the Fiscal Affairs Cham-
            ber of the Court of Appeal of Antwerp." Pages 135–143 in
            *Disputes and the Law,* edited by Maureen Cain and Kálmán
            Kulcsár. Budapest: Akadémiai Kiadó.

Vera Institute of Justice
    1977    *Felony Arrests: Their Prosecution and Disposition in New
            York City's Courts.* New York: Vera Institute of Justice.

Wahrhaftig, Paul
    1982    "An overview of community-oriented citizen dispute resolu-
            tion programs in the United States." Pages 75–97 in *The
            Politics of Informal Justice,* Volume 1: *The American Ex-
            perience,* edited by Richard L. Abel. New York: Academic
            Press.

Wanner, Craig
    1974    "The public ordering of private relations. Part one: initiat-
            ing civil cases in urban trial courts." *Law and Society Re-
            view* 8: 421–440.

    1975    "The public ordering of private relations. Part two: win-
            ning civil court cases." *Law and Society Review* 9: 293–
            306.

Weber, Max
    1925    *Max Weber on Law in Economy and Society,* edited by
            Max Rheinstein. Cambridge: Harvard University Press,
            1954 (second edition; first edition, 1922).

Welsh, David
    1969    "Capital punishment in South Africa." Pages 395–427 in
            *African Penal Systems,* edited by Alan Milner. New York:
            Frederick A. Praeger.

Werthman, Carl, and Irving Piliavin
    1967    "Gang members and the police." Pages 56–98 in *The Po-
            lice: Six Sociological Essays,* edited by David J. Bordua.
            New York: John Wiley.

Westermeyer, Joseph J.
   1973   "Assassination and conflict resolution in Laos." *American Anthropologist* 75: 123–131.

Wheeler, Stanton, Bliss Cartwright, Robert A. Kagan, and Lawrence M. Friedman
   1987   "Do the 'haves' come out ahead? Winning and losing in state supreme courts, 1870–1970." *Law and Society Review* 21: 403–445.

Williams, Linda S.
   1984   "The classic rape: when do victims report?" *Social Problems* 31: 459–467.

Wimberly, Howard
   1973   "Legal evolution: one further step." *American Journal of Sociology* 79: 78–83.

Wiseman, Jacqueline P.
   1970   *Stations of the Lost: The Treatment of Skid Row Alcoholics.* Englewood Cliffs: Prentice-Hall.

Wishman, Seymour
   1981   *Confessions of a Criminal Lawyer.* New York: Penguin Books, 1982.

Witty, Cathie J.
   1980   *Mediation and Society: Conflict Management in Lebanon.* New York: Academic Press.

Wolfgang, Marvin E.
   1958   *Patterns in Criminal Homicide.* New York: John Wiley, 1966.

Woodburn, James
   1979   "Minimal politics: the political organization of the Hadza of North Tanzania." Pages 244–266 in *Politics in Leadership: A Comparative Perspective,* edited by William A. Shack and Perry S. Cohen. Oxford: Clarendon Press.

Yablonsky, Lewis
   1962   *The Violent Gang.* Baltimore: Penguin Books, 1966.

Yngvesson, Barbara
   1976   "Responses to grievance behavior: extended cases in a fishing community." *American Ethnologist* 3: 353–373.

1978 "The Atlantic fishermen." Pages 59–85 in *The Disputing Process—Law in Ten Societies,* edited by Laura Nader and Harry F. Todd, Jr. New York: Columbia University Press.

Yngvesson, Barbara, and Patricia Hennessey
1975 "Small claims, complex disputes: a review of the small claims literature." *Law and Society Review* 9: 219–274.

Youngblood, Gene
1970 *Expanded Cinema.* New York: E. P. Dutton.

Zablocki, Benjamin
1971 *The Joyful Community.* Baltimore: Penguin Books.

# Index

176 INDEX

Social class: 105; of adversaries and law, 21, 69, 102; of criminals, 72, 87–88; of third parties, 15–16, 37; of witnesses, 18. *See also* Occupation and law; Social status

Social control, 54, 120; nonlegal, 78, 79, 83–85, 123–126; quantity of, 123. *See also* Conflict management

Social differentials in the handling of cases. *See* Discrimination

Social distance: between adversaries and law, 8, 29, 34, 59–60; and rules, 92–94; and third parties, 17, 32, 36, 90; and witnesses, 29. *See also* Cultural distance and law; Relational distance

Social diversity, 55, 59–64, 94

Social engineering. *See* Sociological engineering

Social evolution, 45, 51, 56, 66, 83, 99, 103

Social inequality, 82, 125; and rules, 92–94. *See also* Wealth, inequality of

Social information, 13, 64–72, 97, 98, 120–121, 127; decline of, 66–67; management of, by attorneys, 33–36; as a quantitative variable, 64–66. *See also* Desocialization

Social integration: of adversaries and law, 8, 9, 21, 29, 34, 76, 101; of lawyers, 38; and race, 119; of supporters, 13; of victims, 101

Social isolation, 34, 39, 45

"Socialization of law," 120

Social mobility, 87

Social policy: and science, 3, 4; and legal sociology, 4, 89, 102. *See also* Legal reform

Social Security, 52

Social status, 129; of adversaries and law, 8–13, 24–26, 28, 30, 33–35, 37, 41, 59, 60, 64, 76, 89–90, 96, 98, 101, 119; of criminals, 72, 87–88; law and, 10, 11, 108; and race, 108, 127; and speech, 18–

19; of supporters, 8, 13–14, 37, 38, 90; of third parties, 8, 15–16, 32–33, 37, 59, 90, 92–94; of victims, 39, 101; of witnesses, 18–19, 37, 90

Social structure of the case. *See* Cases, social structure of

Societies: agricultural, 74, 115; ancient, 98–101; complex, 59; democratic, 79; lawless, 74, 121; modern, 9, 12, 17, 41, 43–46, 49, 50, 52, 56, 64, 67, 73–75, 84, 85, 87, 88, 93, 97, 110; peasant, 5, 75; primitive, 101; simple, 75, 82, 115, 117; stateless, 74, 79, 80, 82, 121; totalitarian, 78, 79, 124; traditional, 47–49, 53, 64, 76, 126; tribal, 5, 12, 17, 41, 43, 45, 52, 54, 59, 60, 82, 98–100, 110, 113, 117; Western, 108, 115, 126, 127. *See also* individual societies

Sociological anarchism, 81–83, 121

Sociological consciousness, 102–103

Sociological engineering, 71, 81

Sociological impossibility, 55, 74

Sociological jurisprudence, 129

Sociological litigation, 4, 23–40, 66, 95, 96, 113

Sociological model of law, 19–22, 89–91, 95, 108

Sociological precedents, 31

Sociology, 55, 89, 94–97, 100, 103, 130; age of, 4, 102–103, 130; of the case, 4–8, 89; facts and values in, 3–4; of jurisprudence, 91–95; unauthorized practice of, 38–39

Sociology of law. *See* Legal sociology

Somalia, 47–50, 53, 54, 56, 116, 118

South Africa, 59

South America, 64

Soviet Union, 64, 78

Spain, 6

Speech, 17–19, 35, 36, 65, 67, 69, 70, 90

Spitzer, Steven, 128 (n.4)